MERTHYR TYDFIL COLLEGE

23877

KV-201-674

Practical
Pre-School

What *Learning* Looks *Like*...

Knowledge and Understanding of the World

Geography and History

Angela Milner

Contents

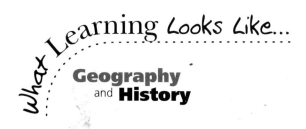
About this book

This book takes a close look at the area of learning which is concerned with children's developing knowledge and understanding of the world. In England, Wales and Scotland this is called Knowledge and Understanding of the World; in Northern Ireland, Knowledge and Appreciation of the Environment. This book focuses specifically on the historical and geographical elements of the area. Another book in this series looks at the science and technology aspects.

Guidance and good practice

The author explains what these elements of the area of learning consist of, what the jargon means and how it applies to the various settings in the Foundation Stage of education - children aged between three and rising six.

There are four different curriculum bodies across the United Kingdom - the Qualifications and Curriculum Authority (QCA) in England; the Curriculum and Assessment Authority for Wales (Awdurdod Cwricwlwm Ac Asesu Cymru); the Scottish Consultative Council on the Curriculum; and the Northern Ireland Council for the Curriculum Examinations and Assessment. Each has a different statement relating to the desired curriculum for young children. However, although there are some differences in terminology and slight variations in emphasis, good practice in one country is still considered good practice in another. The text includes references

throughout to the different systems which allows early years practitioners to put the correct terms in their planning and for inspection purposes.

As well as giving theoretical guidance - albeit in practical terms - the book also aims to give an outline of activities which can be used to deliver the curriculum requirements.

Practical activities

All these activities should be considered as merging into the normal life of the early years setting. The book stresses the importance of play and how the areas of learning are linked. The activities should not be seen as tasks to complete. They are designed to be manageable and fun.

There are suggestions for 15 activities. Some are simple and straightforward activities which need few resources. Others are deliberately more challenging. Each can be used in some form with children at different stages of development.

The activities are planned to cover all aspects of historical and geographical development in the early years. Each one focusses on specific Early Learning Goals (ELGs) but since these overlap and interrelate with one another the activities, therefore, often fulfil more than one goal.

All the activities follow the same format:

❑ The learning objectives are outlined.

❑ Suggestions are made for each activity. The activities cater for different stages of development and involve identification, analysis and evaluation.

❑ Ideas for assessment are given.

❑ Where relevant, suggestions are made for further activities and books and resources linked to the theme.

Children develop at different rates and within the Foundation Stage their progress will not necessarily be steady or uniform. We must not focus, therefore, on age appropriate activities but developmentally appropriate activities. Within your group, you will probably have children working at all three of the suggested activity levels: identification, analysis and evaluation.

These activities are only ideas and should not be seen as prescriptive models. You may want to adapt and change them to make them more appropriate for the group of children you are teaching.

Planning

A planning chart has been included for support and guidance but we would encourage you to adapt this readily to meet your own needs and circumstances.

There are seven books in this series, and although each book can be used by itself, they are designed to fit together so that the whole learning framework is covered.

The seven titles are:

❑ Personal, Social and Emotional Development

❑ Communication, Language and Literacy

❑ Mathematical Development

❑ Knowledge and Understanding of the World: Geography and History

❑ Knowledge and Understanding of the World: Science and Technology

❑ Physical Development

❑ Creative Development

All of these books carry some activities based on common themes which, when used together, will give enough ideas for a cross-curricular topic over a half or even a full term.

The common themes are:

❑ Seasons

❑ Water

❑ Colour *

❑ All about me

❑ People who help us

All the books together provide an outline of the learning which should be taking place in the Foundation Stage.

* Other titles in this series include the theme of 'Colour'. This was not considered appropriate for this area of learning and so rather than suggest contrived activities the author has included an additional activity in the main section.

Assessment

Each activity includes suggestions for assessment. Assessment involves two distinct activities:

❑ The gathering of information about the child's capabilities.

❑ Making a judgement based on this information.

Assessment should not take place in isolation. We assess to meet individual needs and ensure progress. The following ideas may help your assessment to be more effective.

❑ Assessment is a continuous process. It should be systematic to ensure all children are observed on a regular basis.

❑ Assessment should always start with the child. The first steps in providing appropriate provision is by sensitively observing children to identify their learning needs.

❑ Assessment should not take place to see how much the child has learned but to plan appropriately for future activities.

❑ You should be a participant in the assessment process, interacting and communicating with the child.

The main way of assessing the young child is through careful observation.

Observations should:

❑ Record both the positive and negative behaviour shown.

❑ Be long enough to make the child's behaviour meaningful.

❑ Record only what you see and not what you think you have seen or heard.

❑ Be clear - before you begin be sure you know what you want to observe.

❑ Be organised - plan ahead, otherwise it will not happen.

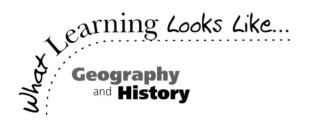

National guidelines

The guidance provided by all four UK curriculum authorities for three- to five-year-olds offers similar, but slightly different, advice on providing appropriate learning experiences for children within this particular area of learning - Knowledge and Understanding of the World.

The Early Learning Goals

The Early Learning Goals produced by the Qualifications and Curriculum Authority (QCA) for England define six areas of learning which form the Foundation Stage for three- to five-year-olds. The six areas are:

❏ Communication, Language and Literacy

❏ Mathematical Development

❏ Physical Development

❏ Creative Development

❏ Personal, Social and Emotional Development

❏ Knowledge and Understanding of the World

The Early Learning Goals (ELGs) establish expectations for the majority of children to achieve by the end of the reception year and provide a basis for long-term planning throughout a Foundation Stage which precedes Key Stage 1 of the National Curriculum. This means that all adults working with children in the Foundation Stage, whatever the setting, must think about

providing a range of appropriate and coherent learning opportunities for the children in their care.

Knowledge and Understanding of the World was first defined in the School Curriculum and Assessment Authority's Desirable Learning Outcomes. It remains a feature of the Early Learning Goals and a key element in the curriculum guidance provided for three- to five-year-olds in other countries within the United Kingdom. This particular area of learning also links closely to the other five areas, providing a curriculum framework to meet the learning needs of all children.

This area of learning focuses on children's developing understanding of the environment, other people and features of the natural and made

world. It provides a foundation for the later development of historical, geographical, scientific and technological thinking and links to the National Curriculum subject areas of history, geography, science and design and technology, which are part of the revised National Curriculum 2000 at the subsequent Key Stages.

Two strands

Knowledge and Understanding of the World has two distinct but complementary strands:

❏ The historical, geographical, human, social and environmental aspects about people, places, space and time, evidence and change, past and present, events and processes;

❏ The scientific and technological aspects of the natural and made world.

There are many links between the statements within the Early Learning Goals and these two particular strands. Many of the processes associated with the development of children's knowledge and understanding in the Foundation Stage can be incorporated into each of the 11 broad expectation statements (see right).

Process not content

The focus in this book is on the first of these aspects - the human, social and environmental aspects about people, places, space and time, the foundation for children's later development of geographical and historical thinking and learning. Within each of the future educational disciplines associated with this area of learning that children will follow at school, there is a strong and distinct tradition. This involves adopting the skills and processes of the discipline itself rather than just learning about the knowledge associated with it. This puts an emphasis on how children learn, not just what they learn, on process rather than content. We should be providing meaningful activities and approaches to support the exploration of two key concepts - the development of a sense of time and a sense of place - rather than worrying about content and the acquisition of knowledge.

Exploring the world

Both of the strands within Knowledge and Understanding of the World use approaches based on:

❑ active learning;

❑ first-hand experience;

❑ sensory approaches.

These approaches promote the development of learning through exploration of the world children inhabit, both inside and outside their early years setting. To meet the

Early Learning Goals for Knowledge and Understanding of the World *By the end of the Foundation Stage, most children will be able to:*	Foundation for later development of curricular thinking
KU 1 Investigate objects and materials by using all of their senses as appropriate	historical, geographical, scientific and technological
KU 2 Find out about, and identify some features of, living things, objects and events they observe	geographical, scientific
KU 3 Look closely at similarities, differences, patterns and change	historical, geographical, scientific and technological
KU 4 Ask questions about why things happen and how things work	historical, geographical, scientific and technological
KU 5 Build and construct with a wide range of objects, selecting appropriate resources and adapting their work where necessary	technological
KU 6 Select the tools and techniques they need to shape, assemble and join the materials they are using	technological
KU 7 Find out about and identify the uses of everyday technology and use information and communication technology and programmable toys to support their learning	historical, geographical, scientific and technological
KU 8 Find out about past and present events in their own lives and in those of their families and other people they know	historical, geographical
KU 9 Observe, find out about and identify features in the place they live and the natural world	geographical, scientific
KU 10 Begin to know about their own cultures and beliefs and those of other people	historical, geographical
KU 11 Find out about their environment and talk about those features they like and dislike	geographical, scientific

national requirements, children will need to experience an environment which offers a wide range of imaginative and enjoyable activities to stimulate their interest and curiosity in the world around them.

There is an emphasis in this area of learning on finding out through direct first-hand experience and observation and discussion, as well as helping children to explore secondary experiences through play. This could be through effective use of existing resources such as toys and games like playmats, or through small world play such as a garage or a dolls' house.

Visits and simple fieldwork in the immediate vicinity of your setting provide excellent opportunities for

active learning. A walk to the local post office to post a letter, a visit to a supermarket to buy a snack or a local museum to discover more about life in the past would all involve children in the exploration of their environment. It makes sense to start this exploration of the world in a context with which children are most familiar - their homes, friends and families, the here and the now. You then need to extend this to include more distant places and times, the there and the then.

Using the senses

A sensory approach introduces children to the world through developing their skills of:

❏ Observation, for example looking carefully at different colours of doors and windows on a locality walk;

❏ Listening, for instance to the sounds they hear in the local street;

❏ Touching, for example artefacts from the past to try to discover what materials they are made from and how they might have been used;

❏ Smelling, for instance to discover whether they like or dislike a particular environment.

There are inherent dangers associated with the sense of taste for very young children when exploring the world they inhabit. Explain that it is not safe to taste unknown substances unless a responsible adult has told them that they can. However, you can include tasting in many of your routine activities, especially cooking. You also play an important role in introducing children to different tastes and flavours, for example in foods from around the world and different cultures. Each child will have different

levels of capability in relation to their sensory development, and you will need to take these into account when planning learning activities and make sure that you offer supplementary experiences for children who have any kind of sensory impairment.

Key questions

Investigations should be planned carefully to ensure that learning is based around children discovering the answers to key questions, to develop their knowledge and understanding of the world and their critical thinking skills.

❏ What is this?

❏ Where is this?

❏ When is this?

❏ Why is this?

❏ How is this?

❏ Who is this?

This gives children the opportunity to solve problems, make decisions, experiment, predict, plan and question. These enquiries lay the essential foundations for developing young children's knowledge and understanding and also start to develop important skills, concepts, values and attitudes.

This area of the curriculum introduces children to a number of important issues and can enhance their language development and their use of a wide range of vocabulary. It also links to mathematical and information and communication technology skills and is essential for the later development of life-long learning skills.

Adults play a key role in scaffolding children's learning and helping them to communicate and record their discoveries. You can provide meaningful and challenging experiences and should encourage children to find appropriate ways of recording their findings. It could be through talk and discussion, through sorting and matching games, encouraging children to make comparisons, engaging them in imaginative and role play or through appropriate cross-curricular vehicles such as model-making.

The Scottish Curriculum Framework

The Scottish Curriculum Framework for Children 3 to 5 identifies planned learning experiences based on five key aspects of children's development and learning:

❏ Emotional, Personal and Social Development

❏ Communication and Language

❏ Knowledge and Understanding of the World

❏ Expressive and Aesthetic Development

❏ Physical Development and Movement

The whole Scottish curriculum framework is designed to help staff plan activities and experiences which promote children's development and learning. It recognises the need for all children to experience a number of features of learning which together form a child's pre-school entitlement. It is different in emphasis from the Foundation Stage for the National Curriculum as recommended in the Early Learning Goals and the Welsh Desirable Outcomes.

Each key aspect is designed to reinforce another. Allowances are made for different types of experience and exploration at different paces to suit individual children's needs is

recommended. Emphasis in Knowledge and Understanding of the World is placed on developing a curriculum to enable children to make sense of the world, using their natural curiosity to investigate environments in a variety of ways. There is a clear desire to move children beyond the world they know and understand and to provide activities that are meaningful and linked to play contexts. The adult's role is to plan an interesting range of opportunities for experimenting and exploring and to provide an appropriate level of support and reassurance. This is achieved through a planning process that provides children with:

❏ Active involvement;

❏ A growing range of skills;

❏ A broadening knowledge;

❏ An extension of their experiences.

In Scotland, the mathematical area of the curriculum is part of Knowledge and Understanding of the World, rather than a separate key aspect linked to providing a foundation for

the National Numeracy Strategy, as it is in England. As a result, there is a greater emphasis on linking mathematical processes to real world uses.

There is also an emphasis on the world of work - children should talk about the routines and jobs of people they meet in the nursery and community, and health education - they should become aware of their own health and safety at home. There is, however, less emphasis in the Scottish guidance on technology, particularly children's ability to work with tools and materials.

The Northern Ireland Curricular Guidance for Pre-School Education

In Northern Ireland, the *Curricular Guidance for Pre-School Education* concentrates on providing guidance for use in a range of settings in the year prior to compulsory education. It defines its curriculum as planned learning experiences appropriate for the age group and is similar to the curriculum received on entry to compulsory schooling. Estimated progression rates are suggested but the guidance acknowledges that children will progress at different rates and that adults should adapt the curriculum and their approach to ensure that children are enabled to achieve.

The curriculum for pre-school education is set out under seven discrete headings, but the expectation is that children will receive their curriculum entitlement in a planned and holistic way through relevant experiences, such as stories, poems and songs. The seven areas are:

❏ Personal, Social and Emotional Development;

❏ Physical Development;

❏ Creative/Aesthetic Development;

❏ Language Development;

❏ Early Mathematical Experiences;

❏ Early Experiences in Science and Technology;

❏ Knowledge and Appreciation of the Environment.

In Northern Ireland, the requisite early geographical and historical learning experiences are separate from the scientific and technological aspects of the curriculum. There is far greater emphasis on the immediate and local

environment rather than on the whole world children inhabit.

The curriculum is based on the provision of appropriate learning opportunities through play and relevant experiences. It emphasises the importance of children, of children's choice in the selection of learning activities and of ensuring that children are valued as individuals. The guidance also outlines the important role of the adult in fostering children's learning. Preparation for the experience children will receive and the links between this and their compulsory education is a common theme throughout the guidance.

The Northern Ireland guidance provides a set of suggested learning experiences. Goals or expected competences are provided for each area of learning under a heading entitled 'Progress learning'. This

provides general descriptors of characteristics and skills the majority of children who have experienced an appropriate pre-school education will display. These statements are closely linked to the 12 statements in Knowledge and Appreciation of the Environment, which provide a range of experiences to help children develop the skills and knowledge required for later historical and geographical experiences at school. There is a particular emphasis on environmental issues, for instance road safety and litter, which are promoted through active work in the immediate environment of the setting.

This guidance provides a narrower and more traditional approach, based on the regular features of early years settings, for example the interest table, the use of play materials, popular topics, the weather and seasons, themselves and their families, stories and rhymes, dressing up and playing, the world of work and simple play based on geographical skills experiences.

The Welsh Desirable Outcomes for Children's Learning before Compulsory School Age

The *Desirable Outcomes for Children's Learning before Compulsory School Age* produced by the Awdurdod Cwricwlwm Ac Asesu Cymru (1996) identify six areas of learning and experiences with Desirable Outcomes within them. They are:

❑ Language, Literacy and Communication Skills;

❑ Personal and Social Development;

❑ Mathematical Development;

❑ Knowledge and Understanding of the World;

❑ Physical Development;

❑ Creative Development.

The guidance provided suggests that provision should be aimed at achieving the Desirable Outcomes. They are not devised as discrete and separate areas of learning but rather as integrated and overlapping. The Welsh advice is not designed as an exhaustive list of competences to record children's achievement against. It recognises that not all children will be able to do all the things suggested and that children with special needs will need alternative strategies to enable them to demonstrate their true capabilities.

Play, first-hand experience, the centrality of the child and the role of adults in challenging children to learn are emphasised through the Welsh guidance, along with the principles of appropriateness and responding to the needs of the individual child. The Desirable Outcomes for Knowledge and Understanding of the World cover both the human and social, and scientific and technological areas as they do in the Early Learning Goals. The guidance suggests a range of experiences young children should receive. These include cultural development, historical development, the world of work and use of money, awareness of the environment and of living things. Enjoyment and confidence are emphasised and should be developed through active learning and investigation.

Certain experiences have to be provided as a requirement. These are experience of other cultures, past events, the work people do, the use of money, the use of the environment, animals and other living things. The guidance also suggests that children should be encouraged to enjoy: pushing, pulling, turning, experimenting, pouring, testing, digging, building and finding out about how things work. The Welsh document claims that these experiences will lay the foundation for the later development of knowledge and skills associated with the curriculum areas of science and technology. Most of the Desirable Outcomes appear to have a distinctly geographical, cultural and general knowledge and skills feel to them. The terminology is similar to that used in the Early Learning Goals but there is less of an emphasis on sensory development.

The table on the following pages indicates the areas of similarity and difference between the curriculum guidance provided by each of the four UK curriculum authorities.

Knowledge and Understanding of the World

Early Learning Goals	Curriculum Framework for Children 3 to 5	Desirable Outcomes for Children's Learning before compulsory school age	Curricular Guidance for Pre-School Education
England	**Scotland**	**Wales**	**Northern Ireland**
Focus: Foundation for scientific, technological, historical and geographical learning	Focus: Key aspects of children's development and learning	Focus: Foundation of confidence in science and technology	Focus: Similar learning experiences to those on entry to compulsory education
By the end of the reception year, most children should be able to: ❑ Investigate objects and materials by using all of their senses as appropriate **KU 1**	In developing their knowledge and understanding of the world, children should learn to: ❑ Develop their powers of observation through using their senses **SKU 1**	By the time that they are five, the nursery experiences that children have had should enable them to:	To help them develop knowledge and understanding of the environment, children should have opportunities individually or in groups to: ❑ Use their senses to explore the immediate inside and outside environment **NIKEA 7**
❑ Find out about and identify some features of living things, objects and events they observe **KU 2**	❑ Recognise objects by sight, sound, touch, smell and taste **SKU 2**	❑ Begin to appreciate the differences in and use of a range of materials **WKU 11**	❑ Explore items on the nature/interest table, for example photographs, plants, natural materials **NIKEA 3**
❑ Look closely at similarities, differences, patterns and change **KU 3**	❑ Sort and categorise things into groups **SKU 5** ❑ Understand some properties of materials **SKU 6**		
❑ Ask questions about why things happen and how things work **KU 4**	❑ Recognise patterns, shapes and colours in the world around them **SKU 4**	❑ Begin to find out about outcomes, problem-solving and decision-making **WKU 7**	❑ Explore with a wide variety of play materials **NIKEA 1**
❑ Build and construct with a wide range of objects, selecting appropriate resources and adapting their work where necessary **KU 5**	❑ Ask questions, experiment, design and make and solve problems **SKU 3**		
❑ Select the tools and techniques they need to shape, assemble and join the materials they are using **KU 6**		❑ Make choices and select materials from a range, exploring their potential, cutting, folding, joining and comparing **WKU 12**	
❑ Find out about and identify the uses of everyday technology and use ICT and programmable toys to support their learning **KU 7**	❑ Become aware of everyday uses of technology and use these appropriately **SKU 9**		
❑ Find out about past and present events in their own lives and those of their families and other people they know **KU 8**	❑ Be aware of daily time sequences and words to describe and measure time **SKU 10** ❑ Be aware of change and its effects on them **SKU 11**	❑ Have a basic understanding of the seasons and their features **WKU 3** ❑ Begin to understand the idea of time: meal times, times of day, sequencing **WKU 4**	❑ Talk about the weather and the seasons at appropriate times during the year **NIKEA 4** ❑ Talk about themselves, for example where they live, members of their family and events in their lives both past and present **NIKEA 5** ❑ Listen to rhymes and stories which have reference to the past **NIKEA 6** ❑ Talk about topics from the children's own experiences, for example holidays, festive seasons and birthdays **NIKEA 2**

Knowledge and Understanding of the World

Early Learning Goals	Curriculum Framework for Children 3 to 5	Desirable Outcomes for Children's Learning before compulsory school age	Curricular Guidance for Pre-School Education
England	**Scotland**	**Wales**	**Northern Ireland**
❏ Observe, find out about and identify features in the place they live and the natural world **KU 9**	❏ Become familiar with the early years settings and places in the local areas **SKU 8**	❏ Talk about home and where they live **WKU 1** ❏ Begin to understand about different places such as the countryside and the town **WKU 2**	❏ Talk about the work of people in the local community **NIKEA 11** ❏ Play with materials associated with different places and occupations, for example seaside, farm, fire station and talk about related ideas with adults **NIKEA 10**
❏ Begin to know about their own cultures and beliefs and those of other people **KU 10**	❏ Understand the routines and jobs of familiar people **SKU 7**	❏ Identify some kinds of work: dentist, doctor, farmer, teacher, postal worker, factory worker, mechanic **WKU 5**	
❏ Find out about their environment and talk about those features they like and dislike **KU 11**	❏ Care for living things **SKU 12** ❏ Develop an appreciation of natural beauty and a sense of wonder about the world **SKU 14**	❏ Appreciate the importance of the environment **WKU 9**	❏ Take some responsibility for caring for their own environment and become aware of environmental issues **NIKEA 12** ❏ Play with simple floor maps and small vehicles, discussing road safety where appropriate **NIKEA 8** ❏ Learn about their pre-school setting and the name of the school to which they will transfer **NIKEA 9**
	❏ Be aware of feeling good and of the importance of hygiene, diet, exercise and personal safety **SKU 13**	❏ Begin to understand about food and where it comes from **WKU 10**	
	❏ Understand and use mathematical processes such as matching, sorting, grouping, counting and measuring **SKU 15** ❏ Apply these processes in solving mathematical problems **SKU 16** ❏ Identify and use numbers up to 10 during play experiences and counting games **SKU 17** ❏ Recognise familiar shapes during play activities **SKU 18** ❏ Use mathematical language appropriate to learning situations **SKU 19**	❏ Have a basic understanding of the purpose and use of money **WKU 6**	
		❏ Begin to understand the use of a variety of information sources, for example books, TV, libraries, IT **WKU 8**	

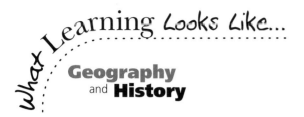

The area of learning explained

This whole area of learning is about reality - the real world, what it looks like, why it is the way it is, how it came to be like this, where and when things are located within it. Children can find out about it by talking to people about their roles and perceptions and through watching its processes as well as through more indirect sources of information such as pictures and books.

The starting point is always with the whole child. Good early years practice and an effective curriculum built on planned experiences will root children's learning in meaningful and

challenging activities, based on their own lives and on the environment with which they are most familiar.

Young children are very self-centred and their experience of the world is likely to focus on themselves, their homes and families and the environments with which they are most familiar - both natural and made. By relating your themes and activities to them, you can help children to move away from an egocentric view of the world to one in which they develop a greater knowledge and understanding, moving from the

familiar and the known to the distant and unknown. Through this approach, children can not only be challenged to learn more about the world and how it operates, but also about their perceptions, viewpoints and misconceptions. This includes developing their awareness of different cultures (KU 10). All of this will contribute towards their development as individual learners.

Sensory approach

The development of a sensory approach - starting with observation then moving on through listening, feeling, smelling and occasionally tasting - is the key to active learning and underpins KU 1. All children, regardless of age and ability, will enjoy and benefit from this sensory approach. Some children will have difficulties if they are unable to use certain senses in the same way that other children can. You will need to plan activities which use alternative senses to compensate.

You can use either direct or indirect experiences. For instance, on a walk around the immediate vicinity of your setting, initial approaches will be based on describing, collecting and interpreting information the children find. This can then be extended to more indirect experiences of different environments and times through the use of toys, books, images, objects and information and communication technology. Children and adults can also find effective ways of recording discoveries through, for example, conversations, displaying

FAR PLACES

PLACES VISITED

PLACES VISITED

FAR PLACES

Talked of places

Work place

Printed places

Recreation

Personal space

Relatives/friends

FAR PLACES

Radio places

Shopping

TV film places

PLACES VISITED

PLACES VISITED

FAR PLACES

objects found, drawings, the use of photographs, model- and map-making.

Young children need help to establish their own sense of identity, their own place within the world and to explore their feelings about the world they share with other people and things. Work covering this area of learning will give them the opportunity to express their feelings about environments and think about their role in caring for and protecting the precious resources we have in our world. Young children will be engaged in learning:

❏ **about** the environment - knowledge;

❏ **through** the environment - as a content vehicle or stimulus;

❏ **for** the environment - green issues.

A questioning approach is vital both in terms of working with children, to find out how their knowledge and understanding is developing, and in terms of structuring learning experiences and curriculum planning.

The spiral curriculum

In planning a sequential and developmental curriculum for young children within this area of learning, it is a good idea to follow a spiral curriculum model, where children are introduced to a number of experiences that they later return to. This progression helps you to cater for different needs and stages of development in the three-stage model suggested: identification, analysis and evaluation.

The spiral curriculum ensures that progression occurs in children's learning. During the Foundation Stage, children are introduced to key ideas and concepts in a simple form that they will meet again later in terms of developing a more sophisticated understanding. This model allows for topics to be revisited and studied in

increasing depth at each stage. Children carry out a range of activities that are increasingly demanding and which take them on a step further in their thinking. Some activities need to consolidate existing learning and others need to allow children to explore new concepts within a known situation. References are made to the spiral curriculum model throughout this book by labelling activities under three stages - identification, analysis and evaluation.

Planning and assessment are closely linked and questioning and observation of activities will provide invaluable assessment opportunities for you to gauge children's levels of competence in moving towards the requirements of the Early Learning Goals. This questioning approach to assessment will also allow you to re-plan and focus activities to meet individual and group needs better. Initially, you will take the lead in formulating the questions that structure children's learning through your planning, assessment and teaching. As children become more engaged with their own active learning, that is the 'how to learn' process, they will begin to ask their own questions, modelled on the teaching and learning experiences they have received.

The curriculum in this area also spirals from the known to the unknown through a progressive sequence of skills and conceptual development linked to the Early Learning Goals. All 11 Knowledge and Understanding of the World Early Learning Goals can be explored in a variety of ways and at a

number of levels, depending on children's previous learning experiences.

The knowledge you gain when observing, monitoring and assessing children's learning will help you to plan for the next stage in a child's development. It is likely that this will be structured through the three-stage approach suggested to allow for continuity and progression in children's development against the national requirements. At first, this will be based around:

1. Identification

❏ Where the children will be involved in observing, handling, listening, naming, describing, labelling, interpreting, spotting similarities and differences.

❏ Where you will be developing children's knowledge and understanding of the world through appropriate learning experiences, questioning the children about their findings, emphasising the use of appropriate vocabulary, increasing the accuracy of observations and helping children develop simple skills in explaining their findings, for example through discussion, listening and talking. Activities may need different approaches depending on whether you're working with three-year-olds just starting nursery, five-year-olds in a reception class, or to match your particular ethos and setting.

2. Analysis

❏ Where the children will be involved in making relationships, sorting, matching, ordering, sequencing, comparing, classifying, making sense of the world.

❏ Where you will continue to develop children's knowledge and understanding of the world through encouraging them to search for pattern and explain more about the things they have explored and how they might be organised or connected.

3. Evaluation

❏ Where the children will be developing investigational skills - using their findings, formulating their own predictions, developing their sense of reasoning, explaining, interpreting and judging.

❏ Where you will deepen children's knowledge and understanding of the world by supporting individual children's ability to communicate their findings and spot the world's interconnectedness as they meet the national requirements within the Early Learning Goals.

This three-stage approach echoes that of the Stepping Stones within the Foundation Stage Curriculum Guidance. The Stepping Stones are designed to help practitioners achieve the necessary progression and challenge in children's learning and

development. Children will progress at their own speed through the identification stage on to the analysis stage and then finally reach the evaluation stage.

The three-stage approach	The Foundation Stage Curriculum Guidance Stepping Stones
Identification	Yellow stage
Analysis	Blue stage
Evaluation	Green stage

The stage children are operating at will relate very much to their previous learning and experiences. If you find that children have any gaps in their knowledge and understanding then you can use the thre stages to reinforce and consolidate what they have already learned. If they are ready to be extended and challenged then you can use some of the ideas to move the children on in their thinking and levels of development to enhance their knowledge and understanding of the world.

The stages of learning

The initial learning stage involves starting with the world that children know well and have experience of as a basis for curriculum activities. You can build on the natural curiosity that children have about places they visit, objects they see and people they meet to extend the type and variety of experiences gained. The first and important stage is a sensitive questioning approach which enables children to look, handle and listen carefully, name, label and interpret evidence, recognising similarities and differences by using their senses to observe and describe their findings. Through asking questions and examining evidence, children will begin to explore in particular Early Learning Goals KU 1, KU 2, KU 8 and KU 9. This initial identification stage is necessary to ensure children can develop more sophisticated levels of thinking. Each new topic should start

with activities designed to develop children's identification skills. Bear in mind that some children will always operate at this level.

The second stage concentrates on extending children's knowledge of and interest in places and people to encourage them to start to make relationships with what they have already seen and experienced. All adults have a role in encouraging children to search for pattern and explanation, to attempt to sort, match, order, sequence and compare. Spotting similarities and differences enables children to classify their findings, locating them in time and space and comparing and associating their findings with existing knowledge and understanding. This enables them to make sense of their investigations and move to a more advanced level of thinking and learning. This analytical stage is particularly important when developing Early Learning Goals KU 3, KU 4 and KU 10.

By the third stage, children, supported by adults, will be using a greater range of investigational skills, evaluating and finding ways of expressing their findings, formulating their own predictions and developing their own sense of reasoning through their investigations. They will provide their own explanations for events they have observed, they will be interpreting evidence and making judgements. This stage is applicable to all the Early Learning Goals from KU 1 to KU 11.

These three stages are essential in enabling children to develop the skills which will allow them to move on to developing their own investigational approaches in their play, lives and learning. They are also the essential building blocks which will help children to connect direct and indirect sources of information and develop their ability to predict what might happen in the world. These skills will help with their decision-making and develop their ability to empathise with others. They will also provide a foundation for recording and recreating their findings using appropriate media and technologies and link closely to the Early Learning Goals KU 5, KU 6, KU 7 and KU 10.

The role of adults

When planning your approach to Knowledge and Understanding of the World, you should think about:

❑ providing practical play activities;

❑ using children's own experience and activities within their own environment;

❑ providing opportunities for children to develop their skills of observation and description;

❑ encouraging an approach which involves investigation and finding out;

❑ developing children's language and vocabulary;

❑ developing children's curiosity in appropriate contexts they can relate to, such as home, your setting, people they know, visits they make.

You should aim to use approaches which enable the children to become mini historians, geographers, scientists, designers and technologists - encourage them to be evidence seekers, travellers, explorers, thinkers and investigators through a process of supported exploration.

Capitalise on children's interests and enthusiasm by providing purposeful fun activities which link in with established themes, resources and activities, for example through stories, rhymes, objects, artefacts, pictures, events and visits.

The progressive nature of experiences and the key role adults play in providing them is also important. Children need challenging, cognitive experiences if they are to acquire a range of appropriate skills and concepts. This will enable them to move from merely describing what they see to comparing their findings and eventually hypothesising why the world is as it is. It will also encourage the development of empathetic skills - children's ability to put themselves in the place of someone else living in a different environment and/or in a different area.

Early geographical learning - developing a sense of place

The study of place is at the heart of developing children's geographical understanding. All young children have a detailed knowledge of the places in their familiar environment. This knowledge is built up through natural curiosity and direct exploration within their immediate environment and through the journeys they make and places they visit or see on television. Children have experience of an increasing number of places as travel becomes easier and more places in the world become accessible to more people. They may have already been on holiday, moved house, collected souvenirs, eaten food and met people from different parts of the world.

You can extend children's knowledge and interest not only of places but also of the relationships between them. Three- to five-year-olds can make simple maps, carry out observations and talk to people about their way of life and the world of work. This is most successful when children are actively engaged in the process of geographical enquiry, when they are motivated and enjoy responding to an investigative approach based on key questions and real issues.

The foundations for geographical learning are laid long before children enter formal schooling, as soon as babies start to move and explore their environment. You can use your own personal geographies and connections, together with experience of more distant places, to help children adopt a geographical way of thinking. Children arriving in your setting will have a wide range of geographical experiences. It is up to you to exploit the fascination with the world that children bring with them to extend their interest. Be careful not to underestimate young children's abilities and their thirst for knowledge and understanding of the world.

Geographical learning experiences should include:

❏ Making close observations of children's surroundings and the people who live and work there;

❏ Going out and about around your building and within your grounds;

❏ Looking at pictures carefully;

❏ Finding out information from a range of appropriate secondary sources;

❏ Recording their findings through drawing, writing, collage, model and map-making;

❏ Looking carefully, handling, listening and interpreting evidence by learning to search for patterns and processes, classifying information and making comparisons between places;

❏ Capitalising on children's interests and enthusiasm.

Imaginative play using toys and construction equipment and role-play environments based around near and far places can be used as effective starting points. Young children can record their findings in a variety of ways including mark making, drawing and modelling. They are also capable of mental mapping. Allowing children to make their own drawings of what they see reveals much about their perceptions of people and places. Attitudes they bring into your setting, for example about other peoples and cultures, can be entrenched and difficult to change. Geographical work helps children to develop a variety of social, physical and intellectual skills, and prepares them for their future life, challenging stereotypes and giving them a greater global awareness.

Useful resources

Many resources can be used to develop children's sense of place and provide appropriate geographical experiences. There is a strong correlation between the resources available in early years settings and the development of geographical understanding. Some ideas are given here (see box, right). You may already have some of these in your setting. You and your colleagues have an important role in choosing and using existing resources to support learning. Geographical work can also be explored through using stories, collections and celebrations as starting points.

You can also make the most of basic equipment and everyday activities to support geographical learning and develop children's thinking in this area, for example:

❏ Outdoor play equipment

❏ Sand and water trays

❏ Mapwork

❏ Large toy play

❏ Information and communication technology

❏ Physical activities

❏ Role-play area

❏ Language and literacy activities

❏ Creative area

❏ Table toys and games

❏ Floor toys

❏ Far-away tables and displays

❏ Large maps, globes, atlases

❏ The weather

❏ Visits and visitors

❏ Construction toys

Resources which promote geographical learning experiences

Pictures, photographs, aerial photographs and postcards of places, weather conditions, land use and geographical features - human and physical

A large base plan of your building and surrounding area

A good set of large photographs of your immediate vicinity and the people who live and work there

A collection of resources based on a contrasting environment or a more distant place

Large scale maps, wall maps, globes, atlases, a large compass

Camera, binoculars, magnifiers

Board games, toys, construction materials, playmats

Containers for collecting

Secondary sources of information such as books, videos, ICT, programmable toys

Dressing-up clothes and role-play props

Weather chart

Sand and water trays

Children can use sand and water trays to create miniature worlds and landscapes, whether from their imagination or to represent a place in terms of its human and physical geography - near or far, wet or dry. They can add props such as rocks, houses, small world people, animals, trees and vehicles to reflect different types of environments and journeys. Children can look at different materials and approaches to building, making underground and overground features such as tunnels and bridges to link places and features. Locks allow children to discover how canal systems work. Water play offers a starting point for exploring concepts such as floating and sinking, clean and dirty water, what happens when we have too much rain or the power and movement of water.

Observing children working with sand and water play equipment will reveal much about the vocabulary of places that they already know and use accurately, such as lake, pond, river. The words children use will link to future geographical terminology associated with landscape features such as hill, river, soil; climate and the weather such as rain, sun and wind; land use such as forest, farm, factory; settlement such as village, town, houses; transport such as car, pedestrian, zebra crossing; and economic and leisure activities such as shops, markets, fairs and parks.

Children could describe a route or journey taken by a play person in their sand or water environment. What vocabulary do they use? Are they aware of how places in their landscape connect to each other? Do they use describing words to describe the direction, position and location of features in their landscapes such as near and far, up and down, left and right?

Jigsaws and board games

Jigsaws and board games can be used as spatial activities to help children to look closely at pictures of different environments and see how things physically fit together. You can use bought puzzles or make your own out of old calendar pictures, cards, magazines or catalogues. The pictures on the puzzles can be chosen to match different countries, times of the day, environments, seasons, and so on, to fit in with your curriculum plan. Encourage children to use locational vocabulary, such as 'Does that piece go above or below that one?' and to think about why an image represents a particular time or place. Questions such as 'How will I get to a certain place?' and 'What will I go past on the way?' can also be explored.

Books

Both fiction and non-fiction books can be used as a way of finding out about people and places and exploring different cultures and traditions.

Stories are set in places and times and often involve journeys or sequencing of events which young children love to predict and follow. A story such as Pat Hutchins' *Rosie's Walk*, which tells the story of a hen's journey across the farmyard, can be recreated with toy models to depict the journey, events and the locational vocabulary involved. Simple informational texts linked to photographs, such as those produced by Dorling Kindersley, can be used to make young children aware that books are a valuable source of information and useful for finding out more about the world. Stories written for young children merge fact and fiction and most children in your setting will already be aware of particular places in televised stories such as Sodor, Greendale and Pontypandy.

Threading, sewing and pegboards

Patterning activities such as threading, sewing and pegboards introduce children to pattern-making and allow them to make or copy images of different places and objects such as boats on water, animals in their environments or recreate patterns associated with particular cultures or peoples. Crafts associated with different peoples, places and cultures can also be explored and made.

Painting, model-making and collages

Painting, model-making and collages can be used to collate children's knowledge of the world. For example, shoe-box models of houses seen on a local walk, paintings of peoples, places and weather conditions, a collage of the three little pigs and their houses of straw, wood and brick, all record children's growing knowledge and understanding of the world and depict sequences and journeys - real and imagined.

Counting skills

Exploration of the world through objects, such as naming the colours of front doors or counting how many cars go past the outdoor play area, will enhance concepts such as change, pattern and busy and quiet as well as improving counting skills in a meaningful context. Children can survey information and present their findings in 2-d or 3-d graphs. This visual communication source can then be used to make comparisons and discover the answers to problems such as 'Which colour front doors are the most popular in our street?'

Environmental print

Environmental sounds and writing in the environment can be used to develop children's language and literacy skills. McDonalds is often the first sign a young child reads. They will recognise words and images associated with particular places and the activities

that take place within them. Children can rub street signs with wax crayons and paper or take photographs of writing they find in the street. You can make tape recordings of particular environments and ask children to link the sounds they hear to pictures of relevant places or create environmental alphabets - for instance r is for road, s is for sign, t is for traffic - to fit in with themes and topics and encourage children to recognise initial letters and sounds.

Role play

Role play enhances children's knowledge and understanding of the activities that take place in particular locations and the world of work. A child can become the lollipop person, stopping the vehicles to allow pedestrians to cross the road in safety. Chalk a road on your playground or outdoor play area and let large wheeled vehicles be the traffic. The children can take it in turns to be pedestrians and drivers. Children can role-play people and events that take place in places they visit - the hairdresser's, the baby clinic, the shops. Hats, costumes and props are essential for children to explore

different aspects of such places. Telephones and computers can be used to communicate with different places associated with the theme.

Songs and rhymes

Action songs and rhymes, singing and music from around the world are a regular feature in early years settings and are all connected with specific places. The wheels on the bus go round and round - but where are the people going? Why do they bump up and down? Where do the five little ducks go to and why? Why do we buy currant buns at the baker's shop? Why do we row the boat? Where do these instruments come from? What are they made of?

Outdoor play areas

Children start to learn about spatial concepts in the outdoor play area. They can look over, through, round, up and down on objects they climb and explore. Outside, children can use large play equipment to recreate the real world and can explore the natural world and seasonal changes. You can carry out projects to enhance the environment such as planting bulbs or collecting litter.

Construction equipment

Construction equipment allows children to build and explore spatial connections and make their own models of miniature worlds and environments. They can make copies of environments they imagine or have seen and use the models they create to guess what things are, work out how places might be used and how the places created are connected to each other.

Playmats

Playmats of different types of environments - urban and rural - can be used to link to work on direction and mapping landscape features. Some playmats are maps of landmasses and are good for locational activities. Try placing holiday souvenirs on a European playmat to work out who went the furthest and which souvenirs came from which countries. Add toy vehicles and props to a locality-based playmat if you're looking at busy and quiet - 'What would happen if a route was blocked?' - or for problem-solving activities such as 'Where should we build a new playground and why?' 'Should we put it near the factory? Near the quarry? Why/why not?'

Environmental walks

Environmental walks and sensory trails are the start of geographical fieldwork. Children can look at photographs of features in their environment and these can be linked to simple adult-drawn maps. They can work out simple problems and identify features of the environment they like and dislike and why. They can describe and record their findings in a whole variety of ways. Items collected en route can be displayed and the weather experienced located on a weather chart.

Information and communications technology

Show children how, with adult support, they can access information through the Internet or through the use of appropriate CD-Roms. Concept keyboards can help young children

Place vocabulary

Physical aspects:

hill	stream	
slope	river	
lake	sea	
waves	land	
soil	sand	
beach	pond	
steep	gentle	
valley	mountain	
wood	forest	
rock	flat	

Settlements:

house
shop
school
park
settlement
village
town
city
building

Transport:

road
car
pedestrian
canal
railway
journey
transport
bridge   tunnel

Locational words:

map	plan
country	
area	place
position	
near	far
left	right
up	down
north	south
east	west

Weather/climate:

year	season
desert	wind
rain	cloud
frost	ice
snow	sun
winter	spring
summer	autumn
weather	storm

Economic activities:

shops	work
jobs	farm
factory	
service	
quarry	
mine	
supermarket	

identify and locate words and pictures associated with particular places. Children will learn to locate keys and use rollerballs or the mouse to move the arrow key or cursor in different directions. Programmable toys such as Roamer can be sent on journeys and simple data can be word-processed or sorted using simple databases such as *Findit* (available from Actis Tel: 0191 3731389).

Matching and sorting

Matching and sorting activities can help children begin to associate animals with particular parts of the world, clothes with different types of

weather, or people with buildings. What do they need to pack in their suitcase for their holidays? How will they get there? What will they need on the way? They can also match 3-d and 2-d base plans of objects, a useful precursor for the development of mapping skills.

Circle time

At circle time, explore children's feelings about places and peoples. Pass around and discuss artefacts from different places. 'What do you think this is?' 'Where do you think this came from?' 'Can we find that place on the map or globe?'

Pictures

Pictures of people and places are the best possible resource if you want to make sure that all children can observe a range of different places and to assess the amount of geographical vocabulary they use accurately.

Visitors

Invite visitors into your setting so that children can listen to people talking about their jobs or the places they have been.

Early historical learning – developing a sense of time

Experts used to tell us that history wasn't an appropriate curriculum area for young children. They believed that children's learning should be based on concrete and direct experiences, that is the here and now rather than the where and when. Yet the past is a natural part of every young child's experience through their own personal histories and contacts, such as parents and grandparents, as well as through television, visits to places of historical interest and the changes they see happening around them all the time. It is just that their understanding of the passage of time occurs in a rather different way to that which we experience as adults.

Stories of famous people and how they lived, noticing whether things are the same or different and experiencing the celebration of events through festivals and anniversaries are all ways of exploring the past and the concept of change. You can help by introducing children to artefacts, talking to older people about life in the past, making visits to museums and historical sites, looking at written evidence of the past and creating appropriate role-play situations.

Young children are curious and willing to ask questions such as 'Who am I? When and where am I?' They are highly motivated and observant of what they see and the physical changes they experience in their own lives. This enthusiasm can be harnessed to allow children to deal with a wide range of historical evidence, if it is appropriately presented through, for example, pictures, artefacts, role play, family history, myths, legends and folk tales, or an exploration of buildings and place names in the locality.

Sequencing skills

Time is a difficult concept for young children to acquire. Adults working in the Foundation Stage should focus on children's sequencing skills, their ability to put things in order, rather than concentrating on measuring time intervals associated with telling the time and mathematics. All children will be aware of time in their daily lives through clocks, birthdays and the changes they experience during the passage of time in a day or events associated with particular days or times of the year. They need to develop a vocabulary of time-related words such as now and then, yesterday and today, morning and evening, and this is best

Key historical vocabulary
then; now
day; night
yesterday; today; tomorrow
past; future
morning; afternoon
breakfast; dinner; tea; supper
week; month; year
last; next; soon
forward; back
long time
clock; hour
while; during
before; after
season; autumn; winter;
spring; summer
date; event; celebration;
birthday; Christmas; holiday
long ago; once upon a time

developed through experiences at home and in their early years setting. Washing lines and pegs are useful for sequencing events in children's own lives - clothes they might wear or have worn, photographs to show how they have grown, or the order routines normally take place in each day.

Stories

Stories are often set in the past and can be useful starting points for an exploration of the past. Children can also construct their own accounts of the past through role play, looking at pictures, mark-making and modelling.

Historical dimension

Popular topics such as 'My family' can be given a historical dimension and explored through a collection of family photographs and by asking questions related to how the family has changed. Young children often spend a great deal of time with their grandparents and talk to them about things that used to exist. This develops concepts associated with old and new, then and now. 'Were you alive when ...?' is a popular question children ask of all adults.

Let's pretend

Young children constantly cross the boundaries between real and imaginary worlds. They enjoy playing 'Let's pretend' and this pretence can take place in a castle, a Victorian kitchen or a pirate ship as well as in the hairdresser's or the home corner.

Objects from the past

History can be approached through a clues approach, with children acting as historical detectives - matching lost objects or bags to people, sequencing objects belonging to different age groups such as a rattle, a push-along toy, a reading book or by sorting new things from old.

Children can handle and describe objects, spot similarities and differences, draw what they see and guess what something was used for. These objects could be borrowed from museums, collected at car boot sales or be replicas of things used in the past that can be bought in museums.

Walks and visits

Use walks around the local area to explain what is old, what is new and how things are changing. Make photographic records to use as a point of comparison. Plan visits to old buildings and museums, which often

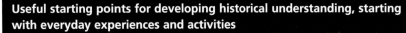

recreate children's lives in the past. What do children remember about their visit and how could it be recreated in your setting?

Events and celebrations

Certain events and birthdays celebrate famous people and allow children to make connections with the past, as with Bonfire Night and Guy Fawkes.

Everyday experiences

Historical development can take place within the context of a number of activities that young children can engage in. These might well be part and parcel of the regular activities offered to the children in your setting.

Everyday experiences and activities which make useful starting points for developing historical understanding are listed (see right).

Useful starting points for developing historical understanding, starting with everyday experiences and activities

Regular events - newstime, sequence of day

Listening to a story or instructions in one part of a session and retelling them later

Sequencing pictures according to time of day, people of different ages (such as baby, toddler, child, teenager, adult), events in a story, seasons, setting day

History far away (ie a long time ago) - What were things like and how do we know?

Memory games such as Kim's game - What has changed?

Matching games such as lotto based on historical themes

Spot the difference - changes over time in photographs

'Guess who?' from evidence bags - Who do the items belong to? What made you think that?

Feely bags - describing the mystery object and guessing what it is

Now and then - for example, looking at photographs of what children can do at different ages/stages of development

Artefacts/objects - What are they? Who would have used them? What for?

History trails - looking for clues/old things on a walk

Drawing, labelling, classifying, collecting old things

Looking at objects/collections children bring from home

Looking at photos: clothes, toys, equipment - What did they look like when they were a baby? What did they have when they were babies? What do they have now? What are the comparisons made and their reasoning for the differences they spot?

Developing children's personal timelines from birth to the present day

Developing appropriate links with clocks

Following events in their family and setting's life through photo albums, diaries, and so on

Learning rhymes associated with time vocabulary, such as 'Solomon Grundy'

Watching video clips, learning songs, dressing up as people from a different era

Talking to older people about when they were little/their age

Visiting museums

Looking at appropriate secondary sources of information with adults, for instance easy-to-read fiction books and CD-Roms

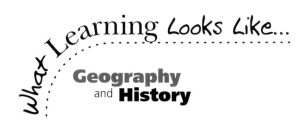

Links with other areas of learning

There are strong links between each area of learning and the Knowledge and Understanding of the World Early Learning Goals.

Communication, Language and Literacy

It would be difficult to cover many of the requirements for Knowledge and Understanding of the World which involve communication about the world without also developing children's language and literacy skills. Speaking, listening, reading and writing all enable children to explore and record their environmental investigations.

The ELGs require children to use language to imagine and recreate roles and experiences of events in the lives of people they know. Role-playing employment situations - teacher, baker, hairdresser - would provide such an opportunity. Using talk to organise their feelings, sequence and clarify thinking, ideas, feelings and events would easily link to children describing their environmental likes and dislikes or discovering more about different peoples and cultures. When they are using their senses to investigate and ask relevant questions about, for example, a box of artefacts from previous eras, they will be listening carefully and responding to what they have heard.

Interacting with others, negotiating plans and activities and taking turns in conversations could all take place during small world play sorting activities. For example, when playing with the dolls' house, you could ask, 'Which of these would we find in a bathroom? A kitchen? A bedroom?'

and so on. Can they work out why an oven is always in the kitchen whereas their toys might be in numerous rooms all over the house? Developing knowledge and understanding of the world requires the development of a specific time and place vocabulary. This is bound to extend their vocabulary, for instance when identifying features with which they are familiar such as in the street outside their setting - 'Is it busy or quiet?' 'What types of vehicles can they see?'

Stories provide an exciting way into activities associated with people, places and times. Children can retell a story they have heard and the sequence of events within it, such as Eric Carle's *The Very Hungry Caterpillar*, where they could sequence the numbers of types of food eaten, the days of the week or explore the life cycle of the butterfly.

In finding out about people and places, children will be using easily accessible non-fiction information to answer their questions with adult support. They will be communicating their findings about objects and peoples they explore by using simple words and sentences, for example to describe members of their family such as mum, sister, myself or their views on their environmental likes and dislikes. They could be involved in designing a poster to encourage others not to drop litter in their own setting. In exploring the world around them, they will see and read print in their environment such as shop signs. Even letter recognition could be explored through a role-play post office where messages for different people could be sorted according to initial letter sounds.

Mathematical Development

Numbers are a natural part of our environment - children know how old they are and celebrate birthdays with cards, badges and candles; they see digital and analogue clocks. They might not know their address but they know the number of their house or how many steps they go up to your setting. As soon as children begin to compare objects or places, they need to draw on comparative mathematical language to describe the differences they see.

Environmental exploration involves looking for similarities and differences and for patterns and shapes. A walk around the locality would allow children to explore the colour, shape and pattern of features such as doors and windows. They could then compare their findings. How many doors were yellow? What shape were the windows? They will play with 3-d objects such as toy buildings which can be drawn around to reveal a 2-d shape to link in to early mapping and they will be using everyday words to describe the position of objects. Where did you put the garage? Was it next to, in front of or behind the house?

Mathematics is part of our daily lives. We count, tell the time, work out quantities and buy objects on a regular basis. Children will experience this in reality when planning a journey or going to the shops. They will also explore these concepts in their play. Role-play areas simulating real life need to reflect a numerical dimension, for example a shop till and scales at the greengrocer's or cheques and credit cards at the travel agent's.

Personal, Social and Emotional Development

In developing children's knowledge and understanding of the world activities will focus on the real world, the environment and people and places. Most children are excited by this area of learning. Children are part of the world they inhabit and they interact with it. Their actions and attitudes can make a difference - they will learn about caring for people and their environment. Children have a strong sense of place and strong feelings about people. They are acutely aware of changes that occur around them such as vandalism to their outdoor equipment and whether this is right or wrong. Encouraging them to be responsible, to become independent and take on jobs in caring for their environment such as picking up litter or watering plants not only develops personal, social and emotional skills but also their knowledge and understanding of the world.

People are an important element in Knowledge and Understanding of the World. Role play provides the opportunity to explore empathetic skills and take on positive roles related to people's jobs. It also allows children to become more independent in terms of dressing, handling equipment and sharing. It can be used to explore a cultural dimension to the curriculum and to emphasise the importance of treating other people's views and needs with respect. People might live in houses made of different materials but they all need shelter. It is important that children think about why and the similarities we all share rather than the differences. Adults working with children must make sure that they don't perpetuate misconceptions by reinforcing negative images or dated stereotypical images of people and places.

Group work, interaction with each other, adults and visitors is not only enjoyable but develops communication skills and ideas about valuing each other's contributions. It encourages children to be confident and to communicate their discoveries, to listen and to concentrate. It helps them to establish good relationships, take turns, share and agree codes of behaviour when, for example, handling photographs or someone's prized possessions. Such small group work is essential

for children to make progress in developing their knowledge and understanding of the world.

Physical Development

In developing their knowledge and understanding of the world, children will be physically exploring objects and their environment. In developing early geographical skills associated with map-making, they will be thinking spatially, showing awareness of themselves and others as they travel and make journeys, play and place things together. Outdoor play equipment allows children to control their bodies but also to explore concepts such as down, up, through, over and round.

Using toys and equipment in a particular setting, particularly construction and small world play, will develop children's fine motor skills. They will be increasing their control as they learn to handle something old or delicate with care and use a range of equipment, such as magnifying glasses to look in more detail at something they have found or noticed.

Creative Development

The environment, people and places are alive with colour, texture and shape. An exploration of model houses along with drawings and pictures would allow children to be operating at a 2- or 3-d level. This mixed approach is an essential experience for children trying to connect objects they see with how they are depicted in pictures and on maps.

As they explore the environment with their senses, they will listen to a wide variety of sounds and learn to associate them with particular places or people. They will also smell, touch and feel. This sensory exploration can then be communicated to others through designing and making to record thoughts and feelings.

Children's creative imaginations will be tapped through the exploration of different cultures, looking at art and design, listening to music, learning dances and listening to stories from another place or time.

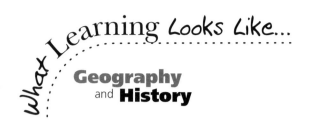

The importance of play

Play is an essential approach to early years education, associated with good early years practice and accepted theories of learning. It is especially valuable within this area of learning, where we want to develop the whole child by challenging them and motivating them to find out more about the world they live in through a range of appropriate experiences.

Play provides a means of initiating, promoting and sustaining learning within the Early Learning Goals. It offers an enabling approach to the early years curriculum, often resulting in high quality teaching and learning experiences. It is also an approach that caters for the needs of all children and enables them to work as individuals as well as collectively, collaboratively and co-operatively. With a play approach, everything children do also becomes observable to adults working with them. This enables more accurate assessment and a greater match of learning experiences for children.

Purposeful play

Play is fun and should always be organised in such a way as to ensure it remains an enjoyable learning experience. However, it needs to be purposeful and structured in order for children to encounter the required learning experiences. Play used in this way is a rich, learning tool with the potential to move children forward in both their thinking and their knowledge and understanding of the world.

Purposeful play promotes learning and allows you to create an enabling

atmosphere for the children you work with. It allows children to learn about the world through supported exploration. It captures their curiosity and encourages them to develop empathetic skills - their ability to put themselves in someone else's shoes.

The Early Learning Goals provide a framework in which you can select methods to enable children to make progress and demonstrate achievement. Through play, children can apply the knowledge they have learned in a more meaningful way to themselves. Play can improve children's language competence and develop key vocabulary and concepts. It can also be used as a context within which children can practise and master skills, allowing them to take on the role of characters and develop their own viewpoints.

Play is a vehicle for learning. Imaginative or small world play, construction equipment, toys and games as well as role-play areas will all provide invaluable learning experiences linked to this area of learning and to the real world environment.

The value of role play

Role play or socio-dramatic play set in authentic contexts and structured by events that children are familiar with and then extended to more distant places and times is also an invaluable learning experience. Role-play areas can become the focus for reconstructions of other environments. Home corners can be turned into places or sites based on visits, events, photographs or stories. They can become a hospital or link to topics on buildings, moving house or the supermarket. Role play should emphasise experiences and exploration. It can be structured, free or self-motivated but the quality of play is dependent upon the imagination of the adults working with the children and the sensitive way in which they observe and respond to children's play. Through play, children can develop their learning in a sequential manner. It is up to you to relate their development to the curriculum goals.

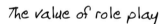

Play approaches to Knowledge and Understanding of the World develop:

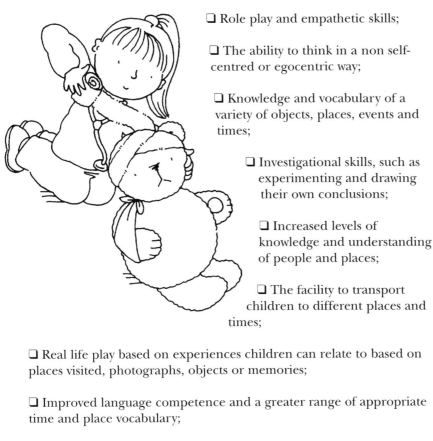

❑ Role play and empathetic skills;

❑ The ability to think in a non self-centred or egocentric way;

❑ Knowledge and vocabulary of a variety of objects, places, events and times;

❑ Investigational skills, such as experimenting and drawing their own conclusions;

❑ Increased levels of knowledge and understanding of people and places;

❑ The facility to transport children to different places and times;

❑ Real life play based on experiences children can relate to based on places visited, photographs, objects or memories;

❑ Improved language competence and a greater range of appropriate time and place vocabulary;

❑ Key concepts such as location - then and now, here and there;

❑ Facilities for practising and mastering skills;

❑ Ways in which activities can be matched to interest levels;

❑ Enhanced understanding of planned sequences of work linked to all the ELGs;

❑ High quality learning;

❑ A character or particular viewpoint;

❑ A differentiated sense of time/place;

❑ The use of their senses;

❑ Skills in handling artefacts, both real objects and replicas.

Play provides situations which allow for resolution of issues by the children.

It allows you to explore topics, stories and events, times and places that you can't provide first-hand experience of.

It enables children to construct and restructure environments they have already investigated, for instance based on video footage of the past or a museum visit.

It offers the opportunity to recreate routines and events that take place at different times of day, in different rooms in a house, or different environments, such as polar regions or tropical rainforests.

It enables children to:

❑ Experience similarity and difference, concepts such as old and new, near and far;

❑ Be involved in problem-solving, for instance treasure hunts, sorting and working out puzzles, such as the odd one out;

❑ Engage in free exploration of objects;

❑ Experience and present drama;

❑ Explore their values and feelings;

❑ Start to develop images, icons, markers associated with time in their minds;

❑ Develop a more accurate picture of life in the past;

❑ Begin to be aware of different viewpoints and their own perceptions.

Where we live

Children learn most from the environment with which they are most familiar. Activities which focus on first-hand experience within the immediate locality of your setting can link to a variety of topics about people and places, for example traffic, the park, shops, the garage, homes and families.

Learning objectives

❑ To help children to establish a sense of place;

❑ To identify similarities and differences between different features and places in the locality;

❑ To understand that people relate to the environment in different ways;

❑ To start to develop skills of first-hand investigation;

❑ To make effective use of the environment as a resource for learning;

❑ To enable children to observe their environment, be curious and ask and answer questions;

❑ To develop a 'where we live' vocabulary.

Curriculum guidelines

Early Learning Goals KU 1, 2, 3, 4, 8, 9, 11

Scottish Curriculum Framework SKU 1, 2, 4, 5, 6, 7, 8, 13

Welsh Desirable Outcomes WKU 1, 2, 5, 9

Northern Ireland Curricular Guidance NIKAE 2, 5, 7, 8, 9, 10, 11, 12

Let's investigate

Identification:
To be active explorers of their locality children need to develop a wide range of investigational skills linked to their senses. Encourage young children to learn to look by asking them what they can see, by playing 'I spy' and hunting or trail games. Children can touch objects and describe surfaces and shapes they feel. Sounds heard in different locations or feelings about places can be explored using a tape recorder.

Analysis:
Encourage children to ask questions about where they live, to collect and classify objects they find, to measure, quantify and record their findings with adult help.

These investigations can take place inside and outside. Outside look at building materials, roofs, walls, the outdoor play area, doors, windows, drainpipes, grates, steps, flags, litter, pollution, the weather and different surfaces. Inside you can look at your setting, signs, people and routines.

Going walkabout

Identification:
The best way to discover more about the immediate environment is to get out and explore it! Local walks can link to the seasons - a winter or spring walk; the weather - rainy day walks, sunny day walks; to ground conditions - muddy wellington walks or barefoot beach walks; to particular themes like colour, shape or buildings or to a journey or route around the locality. Explorations can be made of the natural environment, the built environment and of the 'green' environment.

During such walks children need to explore the environment through using their senses. Small groups of children need to be accompanied by an adult who can focus their observations and use questions to assess what they have discovered.

Analysis:
Early mapwork skills can be developed through focusing on concepts such as direction or location. Think constantly about the language and vocabulary you use with the children. Ask questions such as, 'Shall we go up the hill today?' 'Which road shall we take?' 'What will we go past as we walk from here to the park?' Each feature that you see can be explored and children can share their knowledge of sights of interest as you pass them - a river, bridge, hill, shop. Walks can be planned in advance or discussed on your return. Children can use photographs, simple adult drawn maps and models of the locality to plan or recreate their walks.

Evaluation:
Encourage children to compare places and features within the environment. If you have a local landmark which you can climb up to so that you can look down on the place they are most familiar with, this will help children to see where features are in relation to each other. Finding and pointing games help children to locate features in their environment and to consider how they relate to each other. Children notice things at their own eye level - remember to get them to look up high and low down for clues about what exists overground and underground in their locality.

The intellectual demands made of children can be varied within such a theme and linked to their growing knowledge and skills. Children can use simple cameras to take pictures of the things they like or dislike or to record specific views. What do you see from the school entrance? A particular child's front door? When you stand in one place but turn in different directions? Photographs taken of the children's homes, your setting and key familiar features in the locality can be stuck on boxes and made into a local street model. Children can use puppets or models of themselves to finger walk around their locality and describe what they see. Even very young children can make marks on paper to reveal something about their own mental images of their environment and how the features within it link to each other. Photographs from a local estate agent can be used as part of a 'find the house' game. 'Who can find a house with a red front door/blue garage/round window?'

Your walks will be recreated back in class through the use of play, construction kits, natural materials, playmats, railway lines and road networks. Conventional resources and the local environment together support children's learning. Reality will also be recreated through the children's imaginative play, particularly role play.

Housey housey

Identification:
Some local building contractors will organise visits to new housing developments and these could be made on a regular basis to witness the processes, inside and out, involved in building a house. Turn the role-play area into a building site by providing large bricks, tools, safety helmets, dungarees and boots. If you have an outdoor sandpit this can be a focus for building walls with light concrete bricks and mortar made from sand and water. Colours, patterns and shapes can be explored through the walls children

make. Provide an indoor sand tray with wet sand, toy trucks, bulldozers, ramps and spades to explore foundations, holes and tunnels.

Analysis:
When buildings are constructed many different people are involved. Who does what and why? What tools do they use? What materials are things made from?

Children's own homes are their base and the centre of their lives. They spend much of their time in and around them or visiting other people's homes. They will notice quickly that homes are different - some have gardens and garages, some have more than one floor, different types of windows, different coloured front doors, different street names and numbers outside. Inside, too, there will be differences. Some will have bathrooms upstairs, some down. Some will have rooms with fireplaces, others will have central heating.

Evaluation:
Children need to develop an accurate house and building vocabulary if they are to share their knowledge about where they live. Encourage them to associate different vocabulary with particular houses they know or pictures they see. What is a house, a bungalow, a flat? How would children define the differences? Where are houses located - in streets, roads, avenues? Why do certain roads have particular names? How do we find particular people's houses? What is an address? How is it made up in terms of numbers and words? Can children memorise their own address?

Toy houses and role-play home corners allow children to discover more about their homes. Which furniture is associated with particular rooms in the dolls' house? Can your home corner become different rooms? Can the children play out different activities associated with particular rooms?

Street furniture

Identification:
Children need to notice that we also find objects out in the street which make street life easier, more comfortable or convenient or things which are used to decorate the streets.

Analysis:
What are these things and how do they relate to the life of young children? Street safety can be explored by looking at pedestrians and drivers, communication through letter boxes and phone booths. Items can be found which have connections with the past or with our physical comfort such as bus shelters or benches to sit on.

It is important for children to feel and be safe in the environment where they live. They need to recognise safe places to cross roads and to play, signs associated with danger and places where they shouldn't play because it would not be safe to do so.

Assessment

❑ Talk to children about the places they know and note down all the vocabulary they use to describe places. (identification)

❑ Set up play people and objects associated with them, for example a postperson and a letter box, a fire engine and a firefighter. Can children match the person to the object? (identification)

❑ How accurately do children look, listen and feel? (identification and analysis)

❑ Play a similarity and difference sorting game. Ask children to sort photographs taken in the locality and to explain why they have put certain pictures together. (analysis)

❑ Discuss with children their likes and dislikes about where they live. (evaluation)

Weather forecasting

Evidence of how the weather affects us can be explored through observation, simple recording activities and the development of a weather vocabulary. It is important for children to experience different types of weather. Weather conditions can evoke particular patterns of behaviour or specific fears and emotions. First-hand experience in a secure environment is the best way to explore this.

Learning objectives

❏ To help children to develop their knowledge and understanding of the weather;

❏ To identify similarities and differences, patterns and reasons for different types of weather;

❏ To understand that people relate to the weather in different ways;

❏ To start to develop skills of first-hand investigation;

❏ To make effective use of the environment as a resource for learning;

❏ To enable children to observe their environment, be curious and ask and answer questions;

❏ To develop a weather vocabulary.

Recording the weather

Identification:

Most settings have a weatherboard or chart where children can select an appropriate picture to match that day's

---

Curriculum guidelines

Early Learning Goals KU 1, 2, 3, 4, 5, 9, 10

Scottish Curriculum Framework SKU 1, 2, 3, 4, 5, 6, 7, 10, 13, 14, 18

Welsh Desirable Outcomes WKU 3, 8, 9

Northern Ireland Curricular Guidance NIKAE 1, 2, 3, 4, 5, 6, 7

There is a specific link within the Northern Ireland Knowledge and Appreciation of the Environment statement 4: 'talk about the weather'.

---

weather. If you don't have one you can make one from card. Children can take turns to add weather symbols - a snowman for snow, an umbrella for rain, and so on - or to draw their own weather pictures.

Analysis:
Use the chart to predict the weather for the next day or to look at patterns and sequences of different types of weather.

Evaluation:
Young children won't necessarily be able to use sophisticated equipment, nor will they have the mathematical ability to measure the weather in sophisticated ways. They can, however, watch how quickly a windmill turns to measure the wind speed. They will know something about temperature inside and out and when they have to wear more or fewer clothes. They can

collect rainfall in a plastic container over a number of days and then compare which day had the most rain and which the least or count with a number of the same sized cubes as a measure of how much rain had fallen. They can leave water outside on a cold afternoon and observe whether or not it is frozen the following morning. They can also design and make some equipment to measure the effects of the weather. What they observe and feel will need to be explained so it is important that adults working with young children help children learn how to use an appropriate weather vocabulary.

Weather prediction

Identification:
Many stories and rhymes link to weather lore and predictions. 'Rain, rain, go away, come again another day' is a popular rhyme that clearly relates to children's own experience. Even the youngest children will be aware that if it rains they will have to stay in but if it is fine they can play out. 'The sun has got his hat on' is evocative of the fun we can have when the sun comes out. 'Rock a bye baby' describes what can happen when the wind blows. 'Incey Wincey' tells what happens to the spider when the rain gushes down the drainpipe.

Analysis:
Your local area may be prone to flooding or drought and children will hear adults talking about the weather all the time. Some sayings help us to predict the weather such as 'Red sky in the morning, shepherds' warning'.

Everyday decisions parents and carers make - 'Shall we go in the car or walk?' 'Shall I hang out the washing/mow the grass?' are determined by the weather and children will learn to read the weather signs, such as darker skies as clouds appear may result in rain.

The weather can be explored through all the senses and is often associated with particular seasons and activities. Children are aware of changes in the weather and how that varies during the day. They will be aware of the rain, clouds and sun and will know when it is getting warmer or colder.

Evaluation:
They will see and hear the weather forecast and be aware of extremes in the weather here and abroad and the devastating effect that rapid changes in the weather can have on people's lives. This can take them further in terms of their knowledge and understanding of the world to distant places. Many natural disasters have prompted adults and children to raise money for those in a faraway land badly affected by a particular type of weather condition that we are never likely to experience.

Weather forecasting

Watching the weather person manipulate the weather symbols on a map is a regular feature of our television viewing. You could make your own version by creating a weather studio in the role-play area.

You will need a large playmat or simple map showing the outline of the United Kingdom and the location of your setting. Hang it on the wall at child height or attach it to a pinboard, whiteboard, flannelgraph or magnetic board. Add the compass points - North, East, South and West - to the top, right, bottom and left of the board and have

a box of symbols that children can attach to reflect their weather forecast. Forecasts can be videoed and relayed to others on a television, or looked at through a rectangular box shape made to look like a television. A tape recorder and microphone can be used to listen to weather conditions, make and record sounds to indicate the weather - stamping feet for thunder - or to record their forecast.

Smart clothes, clothes to match particular weather conditions, themed music, a desk, weather pictures, and so on, can all be used to enhance the quality of the play. Satellite images of other parts of the world and cloud formations depicting atmospheric conditions can also be made available. The weather forecast can be set in a TV news studio. Children will need to work together to make and listen to the forecast. They will need to present their forecast to an audience using appropriate vocabulary and ways of speaking. They will have to manipulate symbols around the map placing them in appropriate locations and linking their explanations to their developing mathematical and scientific and technological understanding.

Assessment
❑ Can children recognise different types of weather in discussion? (identification)

❑ What weather word vocabulary do children possess? Note down what they say. (identification)

❑ Can children match clothes they would wear to different weather conditions? (analysis)

❑ Can children accurately record the weather on a weather chart? (evaluation)

Going shopping

The routines associated with visits to shops and recreations of this in children's small world or role play are a natural part of many early years settings. They can also be used to extend children's knowledge and understanding of the world. Shopping is a common experience shared by all children. Some will also have experience of shopping in markets, at fairs, jumble sales and car boot sales.

Learning objectives

❏ To help children to develop their knowledge and understanding of different types of shops;

❏ To identify similarities and differences, patterns and reasons for different types of shops;

❏ To understand that people relate to shops and shopping in different ways for example as employees and customers;

❏ To start to develop skills of first-hand investigation;

❏ To make effective use of the environment as a resource for learning;

❏ To enable children to observe their environment, be curious and ask and answer questions;

❏ To develop a shopping vocabulary.

Identification:

There are various starting points for this theme. It could be an existing role-play shop, a visit to a local store or a story such as Suzanne Garland's *Going Shopping* or Pat Hutchins' *Don't Forget the Bacon*. Many games and factual

books are based on shopping experiences. The Dorling Kindersley book on *Shops* contains pictures for children to sort into different premises. To play the Shopping List game children turn over cards to see if they can find the items on their list to

fill their shopping trolley first. Many routine everyday activities provide appropriate learning experiences to enhance children's knowledge and understanding of the world, but there are other activities you could try too!

The world in a supermarket bag

Analysis:

Arrange a visit to a supermarket and buy all the ingredients to make a fruit salad. When you unpack the shopping together and as you prepare the dish, talk about themes and issues which will develop children's knowledge and understanding of the world. Where in

the supermarket is the fruit located? Is it loose or wrapped? Do you buy it by weight or in pre-weighed packs? What

---

Curriculum guidelines

Early Learning Goals KU 1, 2, 3, 4, 8, 9, 10, 11

Scottish Curriculum Framework SKU 1, 2, 4, 5, 6, 7, 8, 9, 14, 15, 16, 17, 18

Welsh Desirable Outcomes WKU 1, 2, 5, 6, 10

Northern Ireland Curricular Guidance NIKAE 1, 2, 3, 5, 6, 7, 10, 11, 12

---

fruits do children know by name? Shape? Size? Colour? Are there any clues on the fruit about where in the world it comes from? Can they find these places on a large world map or board atlas? Which fruits come from different countries? Are they hot or cold countries? Where does the fruit grow - on trees, bushes, in the open air or under glass, overground or underground? Which bits of the fruit do you eat? What do they taste like? Smell like? Which fruits do children like and why?

Butcher, baker and candlestick maker

Certain professions are associated with particular types of shops. Some popular rhymes are associated with shops and people who worked in them in the past. Others survive today but how many of these professions do children know and link to particular shops? Do children have experience of the butcher and the baker or is it just the counters in the supermarket? Pictures, puzzles and games are available which link to jobs people do and to the world of work. Friezes, books and stories link to these roles as do Happy Families card games and the Allan Ahlberg *Happy Families* series. Objects can be sorted according to who would have made them and which shop sells them. Such an activity would help adults working with children to discover more about their shopping experiences.

Evaluation:
In the past shop signs consisted of pictures of the goods sold. Can children guess the shop from the sign or design a sign for a particular shop? Living museums such as Beamish, Blists' Hill and Wigan Pier as well as many local museums have a display or recreation of shops in the past. Visits to these types of resources can enhance children's knowledge and understanding of life in the past.

The lost shopping bag

Fill some shopping or carrier bags with different objects which can be sorted according to whom they might belong to and which shops the person who has lost the bag might have visited. Their journeys could be recreated using maps, models or toy play. (The Early Learning Centre's Happy Street contains locations for characters to visit including the flower sellers and the garage. Some of these places also contain auditory clues as to what might happen there.)

Inputs and outputs

Many large supermarkets now welcome visits from educational groups. This means you can explore the whole process of what happens to a product from when it arrives in a shop to when it is sold and taken home by the customer. Who puts things where? Which products are put together? Are there any hygiene implications? How are things packaged? What happens to the items they can't sell? How is the packaging recycled? What do people use to carry their shopping home in? Does the supermarket have a recycling centre or skip? Is there a problem with litter or people not taking their trolleys back to the supermarket? The Tidy Britain Group has a number of useful resources including stories, games, posters and puppets that would enable an exploration of green issues associated with this topic.

Converting the role-play area

It is not unusual for early years settings to have a role-play area linked to a shopping theme. Children's knowledge and understanding of the world can be enhanced by making this a quality learning environment, where several of the Early Learning Goals can be planned and assessed through play. Role-play areas provide social settings within which young children will

Useful address
Tidy Britain Group, Elizabeth House, The Pier, Wigan WN3 4EX. Tel: 01942 824620.

develop their personal, social and emotional skills and they should provide a numerate and print rich environment. Shops and services can be recreated - the hairdressers, the grocer, the superstore or the garden centre.

Costumes, hats and props will be required to enhance children's play. A visit to a real shop, which is then recreated by adults and children together, provides the most useful learning experience. Recreations of shops visited in a town centre should reflect what you would find there - clothes, shoes. Props and packaging need to match the location. Props can be made by the children, or a mixture of real, play and replica objects can be used. You will need to structure the environment and encourage the children to take on roles. Different problems, roles or issues can be developed each day to enhance the children's play.

Assessment
❑ Can children recognise different types of shops and the people who work there in pictures and in discussion? (identification)

❑ What shopping vocabulary do children possess? Note down what they say. (identification)

❑ Can children match the clothes certain shop workers would wear? (analysis)

❑ Can children design an advertising poster for a particular shop to show what they sell and why it is a good place to shop for certain things? (evaluation)

Geography and **History**

Outdoor play

Many settings have a permanent outdoor play area with equipment that can be used specifically to allow children to extend their knowledge and understanding of the world. Some of this will be available commercially, some will have been designed to meet the learning needs of your particular children and some of it will be the natural environment. Some activities, such as gardening, normally take place in the outdoor environment. The outdoor play area also provides an opportunity to transport children to different environments, near and far.

Learning objectives

❑ To move about the outdoor environment in safety;

❑ To use language to describe position and location as well as the main features of the outdoor play environment;

❑ To explore the real world through outdoor recreations of places and jobs;

❑ To find out more about the environment of a particular early years setting;

❑ To engage in activities associated with the outside world.

The built environment

Identification:

From your outdoor play area look around your built environment. Ask the children a number of questions to assess their observational skills.

❑ How many buildings can you see?

❑ What are they used for?

❑ What shapes can you see in them?

❑ What are they made of?

❑ Can you see any patterns or colours?

❑ What surfaces can you feel? What are they made of?

❑ Is there anything out of place?

❑ What outdoor equipment can you see?

❑ Is it there all the time or are some things put away?

Looking down

Analysis:

From your outdoor play area ask children to climb up and look down on what they see. Do they see different things? What would a bird see as it flew overhead? Would things look the same from above as they do from the side? Can they match photographs of different objects taken from the side and above or work out where someone might have been standing to take a particular photo?

Round and about

Many outdoor play areas are used for large vehicles and asphalted areas often contain road markings. Markings, toys and cardboard models can be added to enhance the quality of the children's play or to turn it into different places such as a garage with petrol pumps and a car wash or a motorway service station. Road safety messages can be conveyed through the use of cones and traffic lights. Children can take on roles such as the police officer or traffic warden to

---

Curriculum guidelines

Early Learning Goals KU 1, 2, 3, 4, 9, 11

Scottish Curriculum Framework SKU 1, 2, 4, 5, 6, 7, 8, 9, 10, 11, 12, 13, 14, 15, 16, 17, 18

Welsh Desirable Outcomes WKU 9, 11

Northern Ireland Curricular Guidance NIKAE 5, 7, 8, 9, 10, 12

Guidance throughout the UK requires children to gain experience of their immediate setting, inside and out.

---

control the traffic. Roadworks and diversions can be set up to recreate the real traffic world. Children can describe the journeys they have made and these can be mapped on models made from toys inside your setting.

Other locations

Outdoor locations offer space in which children can move. Place previously working objects, such as old rowing boats and oars, there to stimulate imaginative play. Scenery from local amateur dramatic societies can be made secure and used to change the outdoor environment into different places and times. Turn an outdoor sandpit into a beach by adding shells and pebbles. Add sand, rocks, toy boats and lighthouses to water trays to recreate seaside environments.

The natural environment

If you have an outdoor area or, better still, a wildlife area, this offers opportunities for exploring a range of habitats and the creatures who live there. Encourage children to treat these environments with respect and involve them in caring for it.

Tunnels, barrels and hoops

Directional vocabulary such as under and over can be explored through the use of tunnels, barrels and hoops. Children can become trains going through tunnels or an animal hiding in its burrow.

Slides, swings and climbing frames

Children will gain different perspectives on their environment when they play on large play equipment. Observe their play to assess the use of geographical vocabulary associated with position and location. Ask them to move along the logs, over the A frame, through a pipe, stand under a bridge and climb up a ladder. You can then assess which vocabulary children were able to recognise and respond to.

Going camping

Evaluation:

Picnics and camping are fun activities that take place outside. Involve children in planning for these experiences as well as recreating them for themselves or toys. Going on holiday or camping would allow children to share responsibilities, engage in teamwork, follow instructions, manipulate objects, engage in imaginative play and explore scientific and technological concepts.

To recreate this experience you need to provide: a tent and ground sheet, sleeping bags, rucksacks, torches, pots and pans, first aid kit, cutlery, plates and cups, lanterns, table and chairs, a washing-up bowl and bucket, clothes and wellingtons, maps and guidebooks, a compass and address book, campsite guide, food and money.

Assessment

❑ Do children know basic road safety rules? (identification)

❑ Can the children describe and direct someone around their outdoor play area? (identification)

❑ Can the children use language or follow directions to describe and move to specific positions and locations? (analysis)

❑ Can the children use the outdoor environment to recreate journeys made or roles people have, such as delivering the milk? (evaluation)

Children can use the equipment provided to plan their camping trip. What would they take with them and why? Where would be a good place to pitch their tent and why? Where would they get their water from? Get washed? Go to the toilet?

What a picture!

Young children can explore the wider world and different times and places through visual materials. Picture reading is an important skill that young children need to acquire. You will need to help them to develop this skill and the necessary vocabulary to ensure that they can explain their geographical or historical thinking.

Every early years setting needs access to a variety of large high-quality pictures of different people, places and environments. These are commercially available but it is useful if you can organise parents and carers to collect suitable pictures, posters, calendars, and so on, which can be used. Large, clear pictures linked to common themes are the most useful as well as pictures that emphasise comparisons or change and illustrate famous people or events. It is important that images are up-to-date and non-stereotyped.

Collect photographs as well as pictures that are other people's representations of reality, such as an illustration in a book. Try to include black and white as well as colour images. Many young children never see black and white images and have misconceptions about the past and how old things are - many mistakenly believe that black and white pictures must be old.

Learning objectives

❑ To enable children to read pictures for information;

❑ To make children aware of different peoples and places through the use of appropriate visual images.

Using photographs of people and places

Snap
Identification:

Use duplicate pictures of people and places to make snap games for children to play.

Sorting and matching
Analysis:

Collect a group of pictures of similar things, such as buildings, for children to sort and match. For example, can they group shops, houses or public buildings together?

The 'Where's Wally?' approach
Identification:

The *Where's Wally?* books require children to look carefully at crowded pictures to find a particular character. You can take a similar approach to any picture, challenging children to find a particular thing or person within it. Can you find so many things that are blue? Are alive? Start with a particular letter?

Story sequence pictures
Analysis:

When you have read or told a story do you let children retell it? How do you do this? With props, puppets or actions? Through pictures is another useful way. At the end of an activity, pictures from the story can be revealed. Ask children to remember what part of the story the picture was from. Was it at the beginning, middle or end? Alternatively ask them to sequence the pictures along the floor, a table surface or a washing line to help them retell the story.

Guess who?
Evaluation:

Ask everyone (including staff!) to bring in their baby photos and try to match them to the children (or adults). What characteristics do they use to work out who is who?

Spot the object
Identification:

Do children know what objects are? Can they recognise pictures of objects taken from peculiar places or in odd places?

Odd one out
Analysis:

Which pictures are the odd ones out and why? This idea can be linked to numerous types of activity, such as 'Which of these things wouldn't be found in a particular place?'

Snapshot
Identification:

Do you have a camera children can use? Can they take pictures of features in their environment? Do you have a digital camera that can be used to record children's observations and produce books and newsletters for parents and other children?

Photo starters
Identification:

Looking at a photograph or a picture can be the start of any discussion. Ask questions related to what is happening.

Curriculum guidelines

Early Learning Goals KU 3, 4, 7, 8, 9, 10

Scottish Curriculum Framework SKU 1, 2, 4, 5, 7, 8, 9, 13, 14

Welsh Desirable Outcomes WKU 2, 8

Northern Ireland Curricular Guidance NIKAE 2, 3, 5

The Welsh Desirable Outcomes specifically require children to: 'begin to understand the use of a variety of information sources such as books, television, libraries, information technology'.

For such young children it will be the visual images associated with these sources that will be particularly accessible. You will need to introduce children to a variety of activities which will enhance children's abilities to access the wealth of visual information available about the world, past and present.

'What can you see/point to/ find?' 'What is that character saying?' 'What might have happened next?'

Where was this picture taken?
Analysis:

Can children match photographs to pictures or maps of locations? What clues do they use?

Pictures of faraway places
Evaluation:

Pictures can be used to introduce children to distant places. In a role-play travel agents children could use the pictures in brochures to choose a holiday destination and explain why they would choose this particular place. Is it like where they live or is it very different?

Complete the pictures

Enlarge photographs of people, objects and places that children are familiar with using a colour photocopier and cut them up into jigsaw shapes. Children can either make the complete picture or keep some pieces and ask children to predict or describe the missing piece they need to complete the picture.

Exhibition

Children can make an exhibition of pictures that are the same and put a frieze together by cutting, sticking and mounting the pictures.

Illustrated alphabet

Give children pictures of things associated with your topic and see if they can produce a word book of things that start with a particular letter to form their own glossary for the topic.

Make a map

Whenever you go on a visit, take a camera with you and use it. Afterwards, children can try to place photographs on the map in the particular location in which they were taken.

Sorting postcards

Children can sort postcards into places near and far, hot or cold or associated with particular countries or landscape features such as the seaside or the mountains.

What can you find?

Can children look at a blown-up or aerial photograph and work out what important things they can see? The youngest children can tell someone or record their answers on a tape recorder. Older children could write a list of what they see using emergent writing or a word bank of pre-prepared words they could match to the picture.

Before and after pictures

Can children draw what might happen next or what has already happened?

The family album
Identification:

Children can look at pictures of their own and other people's families to investigate life in the past. What did people wear? What are they doing? Where are they? How old do you think they are?

The moving image
Analysis:

Children are confronted daily with moving images of life today, in their country and in different parts of the world. They can use these images to compare places they know with places far away. Similarly, the past can be explored through video images which transport children back to distant times and the more recent past when parents and grandparents were young.

Assessment

❑ Can children read pictures for information? Note down what they observe and the vocabulary they use. (identification)

❑ Are children aware of different peoples and places? Do they have any misconceptions based on stereotyped images of peoples, places and other cultures? (analysis/evaluation)

Let's have a celebration!

Children love routine and their lives are marked by key events associated with rites of passage and the ages and stages in their lives. Three- to five-year-olds will have already accumulated a number of firsts in their own lives which will have been celebrated by their families, such as their first smile, first tooth, first steps, first word and first birthday. Similarly, each year has its own pattern, with celebrations marking birthdays, public holidays, celebrations of annual and special events associated with our cultural heritage, as well as local, national and international events. Birthdays, festivals, anniversaries and centenaries all provide motivating and memorable learning experiences for children and exciting ways into studying life in the past.

Children will associate these events with certain times of the year and with particular people as well as celebrations. They will mark the passage of time for children who will associate events with 'When I was three, four or five'. Their birthdays are key events in their lives, celebrated in both their homes and their early years setting. But birthdays, cards and presents are not universal and will not be celebrated in the same way by everybody. Similarly, different religions celebrate different festivals and rites of passage as children grow into adults.

Learning objectives

❏ To understand more about the nature of particular celebrations and what they commemorate;

❏ To find out more about the past and

present and how they link to their own lives and families;

❏ To begin to learn more about other people's cultures and beliefs.

Famous people

Identification:

Events we celebrate often commemorate the lives of famous people, real or imagined. Young children are fascinated by stories of famous people and often want to find out more about how they lived their lives in the past. Pictures or photos of famous people can be used to explore the past. What did they wear? Where did they live? When did they live? What did they eat? Where did they buy things? How did they travel? What did they do at work/in their spare time?

Tell stories of famous people to help children to understand why these people became so famous. They can recreate their stories through their play or by retelling them with the aid of props or pictures of particular objects associated with that person. Play 'Guess the famous person' from props in an evidence bag. For example, Father Christmas might have a reindeer, a letter and a parcel. Guy Fawkes might have a barrel, a date marked in a calendar of 5 November and a picture of the Houses of Parliament.

Celebration vocabulary

Each celebration will have its own vocabulary associated with particular events. If children can describe such events they will explore vocabulary associated with time conventions as well as particular people and events.

The key vocabulary to encourage children to use is:

> then; now
> day; night
> yesterday; today; tomorrow
> past; future
> morning; afternoon
> breakfast; dinner; tea; supper
> week; month; year
> last; next; soon
> forward; back
> long time
> clock; hour
> while; during
> before; after
> season; autumn; winter; spring; summer
> date; event; celebration; birthday; Christmas; holiday
> long ago; once upon a time

Analysis:

Children can talk about celebrations they are involved in to commemorate events and links with the past.

Stories exploring celebrations and events

Stories can be shared with children to enable them to find out more about how different people celebrate anniversaries and festivals. Particularly useful are *PB Bear's Birthday* video and CD-Rom, Jane Hissey's *Ruff* - the story of a dog who doesn't know his age so all the toys welcome him to their family by celebrating a birthday a day for a week with appropriate presents. Shirley Hughes' *Lucy and Tom's Christmas* and E Lloyd's *Nina at the Carnival* allow children to see how other people celebrate different events.

A calendar of celebrations

Evaluation:

When planning events in your setting over a term or year, there will be something to celebrate during most weeks and so you will need to take care with dates and provide a balance of celebrations to ensure they reflect the diversity of life in the United Kingdom. Some religious groups will not want their children to be involved in particular celebrations and parents should be informed about your setting's policy. Celebrating an event, be it Pancake Day or Hallowe'en, will involve the children in memorable and enjoyable activities which are associated with an event or people being commemorated. Children will make cards, decorations, food and presents to give to others as part of their celebrations. Photographs of what they have done and how they have celebrated these events can be taken as a record and displayed to show how your setting has celebrated a year in its life.

Curriculum guidelines

Early Learning Goals KU 2, 3, 8, 10

Scottish Curriculum Framework SKU 1, 5, 10, 18

Welsh Desirable Outcomes WKU 1, 3, 4, 8

Northern Ireland Curricular Guidance NIKAE 2, 3, 5, 6

There is specific reference within the Northern Ireland Knowledge and Appreciation of the Environment to encouraging children to:

'talk about topics which arise naturally from the children's own experience, for example, holidays, festive seasons and birthdays'.

Celebrating the children in your setting

Every child brings their own unique history and values to your setting. They have their own rules to games and words for familiar objects based on their family upbringing and their cultural background. Everyone working with young children wants to value and enhance what children bring but also help them to develop a strong sense of time and to establish their role as a unique, individual part of history.

Activities that help children to establish a sense of time and order are essential. Tidy-up time, going home time, and so on, will all help to convey a sense of when and how long things take. You can time activities, such as how long it takes to bake a cake, and share the numbers and movements of the clock hands with children to mark the passage of time. Key words should be used in natural conversation so that they become part of the children's vocabulary.

The diversity in each setting should be valued and opportunities developed to recognise and value different cultural backgrounds. There are many overlaps with the personal, social and emotional development of children.

Celebrating a festival

Your approach to celebrating different religious festivals should be sensitive to the religious background and commitments of children's families and the context of your own setting.

Learning about a festival and how we celebrate it will involve teaching children about features of a particular religion that may be new to them or, if familiar, could help children to understand their own cultural and religious background. Work on festivals should focus on:

- adopting a multi-sensory approach
- opportunities for structured play
- using artefacts
- visiting the local environment or having knowledgeable visitors into your setting to explain more about the festival.

It is important that the way events are celebrated doesn't undervalue or stereotype particular religions. One way many settings handle this is by looking for similarities rather than differences. Looking at festivals of light, for example, would allow children to explore Advent and Christmas (Christian), Hanukkah (Jewish) and Diwali (Hindu).

Settings in Scotland, Wales and Northern Ireland will celebrate particular national events and the national days associated with their patron saints. Stories, names and emblems can be explored to help give meaning to the events used to commemorate these famous people and events.

Assessment

- What celebrations do children know? How are they celebrated? (identification)
- How do the children celebrate events at home? Which events do they celebrate and why? (analysis)
- Talk to the children about celebrations they have been involved in. What did they enjoy about them? What did they remember? Why did they celebrate them? What did they learn about other peoples' cultures and beliefs? (evaluation)

Time travellers: visiting a museum

A museum visit enables children to become time travellers to discover more about life in the past. Museums offer a wealth of opportunities, experiences and resources. They provide:

❏ A source of information

❏ An opportunity for children to find out more

❏ Motivation and inspiration which captures a child's imagination

❏ Real artefacts

❏ Emotional and empathetic experiences

❏ An appreciation of life in the past

❏ Special exhibitions linked to topics, events and celebrations

It is worthwhile contacting your local museum service to become more aware of the opportunities they provide and how you can use their facilities effectively.

Learning objectives

❏ To develop children's interest in life in the past;

❏ To develop children's knowledge and understanding of how people used to live;

❏ To develop children's observational skills through using their senses,

looking for patterns and relationships, similarities, differences and sequences;

❏ To nurture children's ability to communicate their understanding of past times to others through appropriate activities such as drawing, painting, modelling, drama, music, movement, handling objects and play;

❏ To develop an appropriate time vocabulary associated with the museum visit.

It is likely that this will be the first visit

to a museum for many in your group. Careful preparation will be required

Curriculum guidelines

Early Learning Goals KU 1, 2, 3, 4, 7, 8, 10

Scottish Curriculum Framework SKU 1, 2, 3, 4, 5, 6, 7, 9, 10

Welsh Desirable Outcomes WKU 1, 4, 7, 8

Northern Ireland Curricular Guidance NIKAE 1, 2, 3, 5, 6, 7

and you will need to ensure you have:

❏ Gained permission to take the children out of their setting;

❏ Made appropriate transport arrangements;

❏ Located the toilet facilities;

❏ Thought about safe places to cross roads;

❏ Thought about what children and adults will do in the museum;

❏ Arranged an appropriate staffing ratio;

❏ Prepared the children and adults for your visit;

❏ Made arrangements with the museum education officer;

❏ Considered which sources/resources

you can buy, loan or borrow to link with your visit.

Before and after the visit

Identification:

Explain to the children that you are going/ have been to a museum to discover more about how things were done in the past.

Allow the children to explore old and new through sharing some artefacts associated with your visit. Many museums can lend you resource collections for this exercise. What were the objects used for? Can the children use them appropriately in their play?

Analysis:

Can they draw them or sort them into objects associated with particular activities, for example cooking, cleaning, washing.

Evaluation:

Emphasise sequencing skills, linguistic skills and children's appreciation of life in the past. Think about how what you introduce to the children links to their existing experience and understanding. How well do they work with others and can they make decisions about life in the past? Were you surprised by the children's responses?

Recreate the theme associated with your museum visit by converting your home corner into an area associated with life in the past. Add a mixture of real and replica objects and costumes to enable children to engage in role play about their visit. Involve adults in helping children understand the sequences and routines involved with life in the past.

During the visit

Repeated visits to the same local museum are ideal and offer many educational opportunities linked to all the areas of learning within the Early Learning Goals. They will also allow children to imagine that they are stepping back in time to discover more about life in the past. You can make a time machine with a giant clock. Turn the clock hands to different times, illustrated with pictures or artefacts associated with them.

Any visit should be fun, so give children the chance to explore, play with and handle artefacts, observe and ask questions. Some activities can be organised for you at the museum, others can take place in preparation for and after your visit. Some visits will involve dressing up and recreating daily routines associated with life in the past. It is best to focus on one display or room and link it to a concept and vocabulary you wish the children to acquire such as 'then and now' or 'old and new'.

Play observational games such as spot the difference, I spy or odd one out. Take photos or video your visit and then recreate part of the museum back in your setting as a role-play environment for children to explore and learn in. Activities should concentrate on using the senses and simple problem-solving. Make a class book of photos, drawings and sayings after the event to share with other groups, parents and carers.

Some museums have demonstrators, facilitators or actors who help to make the past come alive and these are particularly valuable. They can show children how water was pulled from a well, the way to the outside toilet or involve them in baking or wash day activities that existed in the past. The best museums are ones which have rooms where children can engage in play activities associated with life in the past and display cases which young children can actually see into.

Using sites and buildings

Young children are increasingly welcomed at historic sites and monuments. Many arrange events such as drama, dance and storytelling with the youngest children in mind.

Assessment

❑ Note down the time vocabulary children used. (identification)

❑ Did the museum visit arouse children's interest and curiosity? (analysis and evaluation)

❑ Did the children ask questions and were they able to share their findings? (analysis)

❑ Did children act out experiences on their return to your setting in formal activities or in spontaneous play? (evaluation)

Useful book

Detailed guidance about planning visits is available in an HMI book *Museums and the Curriculum* (1988) HMSO.

Useful website

The English Heritage website can keep you up-to-date with the latest information and events:

www.english-heritage.org.uk

Educational services contact: 01793 41 49 10

Themes based on periods (for example Romans), buildings (such as castles) or lives of famous people associated with particular places, can be explored and useful information is available via the Internet. Sites often contain landscape models and tableaux of life in the past which help to make history come alive.

Work based around a visit to a site or building would form an exciting focus for work across the curriculum as well as allowing the recreation of the site back in your setting. Lots of experiences like this provide children with a background knowledge on which to build their historical understanding.

ACTIVITY

The way we were

Children are fascinated by the changes they see going on around them and particularly changes that have occurred in their own lives. 'The way we were' can be used as a topic to look at how children have changed and explore their developing sense of time. You can then extend this further back in time to enable children to find out more about the past and present events not only in their own lives but in those of their families and other people they know. This will help children to become historical detectives, learning to interpret evidence to discover more about life in the past.

Learning objectives

❏ To develop children's interest in 'the way they were';

❏ To develop children's knowledge and understanding of how people change;

❏ To develop children's observational skills through using their senses, looking for patterns and relationships, similarities, differences and sequences;

❏ To facilitate children's ability to communicate their understanding of past and present times to others through appropriate activities such as drawing, painting, modelling, drama, music, movement, handling objects and play;

❏ To develop an appropriate time vocabulary associated with the way they were.

Now I am three, four or five

Identification:

Children can compile a book or make

Curriculum guidelines

Early Learning Goals KU 1, 2, 3, 4, 7, 8, 10

Scottish Curriculum Framework SKU 1, 2, 3, 4, 5, 6, 7, 9, 10

Welsh Desirable Outcomes WKU 1, 4, 7, 8

Northern Ireland Curricular Guidance NIKAE 1, 2, 3, 5, 6, 7

There is a specific historical dimension to the English Learning Goals where children are expected to:

'find out about past and present events in their own lives, and in those of their families and other people they know'.

Children will be involved in communicating their findings, thinking, using evidence, recording, investigating and appreciating life in the past.

---

a display to show how they each celebrate their birthdays. It could include photos, birthday cards, badges and presents. You could talk about birthdays at sharing or circle time.

Analysis:
Explore evidence of how they have changed through photographs brought in from home. Ask them to talk to parents or carers to discover more about what they could do when they were a certain age or how old they were when they first smiled or walked.

My life

Ask children to bring in pictures of themselves at different ages - if possible at ages one, two, three, four and five.

See if they can sequence the pictures and make a zig-zag book to show how they have changed over time. You could create a large frieze showing the whole group.

Evaluation:
Talk to the children about their memories and find out if they can remember some of the things that exist in the picture evidence, such as favourite toys. This activity will enable children to start to build up their own personal history. It will also help them to explore appropriate vocabulary through discussion with each other and adults. For example, which is the earliest/latest/most recent photograph? Why do they think that? What clues exist in the picture that could help them to answer these questions?

My personal timeline

Identification:
Use markers or pegs to define ages on a personal timeline. Children can then use this to sort pictures or artefacts according to what they would have worn or played with when they were a particular age. Baby clothes, shoes, toys or photographs from children's clothes catalogues are all useful for this.

When I was little

Analysis:
Children will be keen to share mementoes and photographs of themselves when they were little and to tell you and their friends what they could do. Play simple lotto or sorting and matching games to help children differentiate between things they could do when they were a baby, a toddler, a child. When did they wear a nappy? When did they sit in a high chair?

10 ways to find out about 'The way we were' with young children

Help children to:

❑ develop a time vocabulary (language and conventions);

❑ understand themselves in time;

❑ discover more about their family history;

❑ recognise that older people are an invaluable source of information about the past;

❑ appreciate that their neighbourhood contains many clues to the past and is subject to change;

❑ appreciate that their early years setting has its own particular history;

❑ appreciate the differences between family life now and in the past;

❑ appreciate and learn something about the lives of some famous people they can relate to;

❑ appreciate that historical events are often celebrated;

❑ begin to appreciate what the way of life would have been like for people living long ago;

All these experiences should form part of children's early historical experiences and activities.

When did they ride a bike? Start nursery or playgroup?

Make a wall frieze associated with the different stages children have gone through. Include the headings: When I was a baby; When I was a toddler; Now I am a child. Attach one end of a length of string to a pointer or a picture of themselves and the other to the wall below the frieze so that children can move along the frieze and describe what they were like when they were little. What do they notice about other people?

My family

Families consist of different generations - children, teenagers, parents, grandparents. Can children

sort out and match the generations to the activities they might like to do? Who would a skateboard, skipping rope or rattle belong to?

Share a story

There are many stories about babies and how they grow, such as *Avocado Baby* or *Miss Brick the Builder's Baby*. Stories encourage children to listen and share experiences. Can children relate to these stories? Do they think they are true?

Try sharing these stories with your children:

❑ *The Baby*, John Burningham (Cape);

❑ *101 Things to do with a Baby*, Jan Ormerod (Picture Puffin);

❑ *The Baby Catalogue*, Allan Ahlberg (Picture Puffin);

❑ *Nancy No Size*, Hoffman and Northway (Little Mammoth);

❑ *Meg and Mog's Birthday Book*, Nicoll and Pienkowski (Picture Puffin).

What happens in each story? Can the children sequence the events associated with the story?

Old Bear

By looking at the way we were, children will be exploring concepts associated with now and then, old and new. In Jane Hissey's *Old Bear*, the toys remember that someone else used to share their life in the toy room - an old bear who was placed up in the attic so that the children wouldn't damage him. The toys embark on an adventure to try to rescue Old Bear and return him to live with them in the toy room. This story alone could be used to plan a range of historical learning experiences:

❑ Talk about the attic - What old things can you see? What were they used for (for example, the carpet beater)?

❑ Why was he called Old Bear? How have bears changed over time? Create or visit a Teddy Bear museum.

❑ What do we do with old things?

❑ Retell the story, sequencing the events that happened or from the point of view of the different characters.

❑ Why are bears favourite toys? What are the children's favourite toys? What were mum and dad's, gran and grandad's favourites?

From one common starting point, work can be differentiated into the three stages:

❑ **Identification:** What happened in the story? What were the old objects in the attic?

❑ **Analysis:** Do you think the story was true? Why? Why not? What were the old objects in the attic used for in the past? What would we use today instead?

❑ **Evaluation:** Why was the main character in the story called Old Bear? What characterises old things? How can you tell they are old?

Assessment

❑ Can children identify three changes that have occurred to them in their lifetime? (identification)

❑ Record how children are able to communicate their findings about the past and the vocabulary they use. (identification and analysis)

❑ Can children sequence three pictures or artefacts associated with their life? Can they sort the lives of other members of their families? (analysis)

❑ Talk to the children to see whether they are curious about the way they were. (analysis and evaluation)

Any old iron

Young children are avid collectors and will often arrive at your setting with objects they have collected. They love to handle and sort objects and you can capitalise on these interests when developing children's Knowledge and Understanding of the World. Starting with a collection will enable children to act as geographical, environmental or historical detectives. They will be able to discover more about the world and how the artefacts they explore fit into it.

Learning objectives

❑ To discover more about the world from artefacts and objects;

❑ To learn how to handle objects with care;

❑ To begin to make connections between objects;

❑ To develop a time and place vocabulary that enables children to explain what they have learned about the world from their collections.

Using artefacts to develop a sense of time

Identification:
Artefacts are objects surviving from the past which can be used to show children what life was like in another time. They are a useful source of evidence that even the youngest children can handle.

It is a good idea to try to build up collections of artefacts to support the themes you regularly cover. Artefacts associated with family life, jobs, shops, holidays, school, home, clothes, food, games, toys and methods of transport would all be useful. Scour jumble and

car boot sales to see what you can find or buy replicas: some museums offer a loan service; many sell replicas.

Pattern, shape, similarity and difference are all concepts that can be explored by working with artefacts. Groups of children will have different opinions about what things are and what they were used for. They will learn that some objects suggest or mean different things to different people. They will also learn that they need to handle artefacts carefully. Artefacts that are unbreakable or not precious can be used in displays that children can handle, in role-play areas or in a class museum.

Curriculum guidelines
Early Learning Goals KU 1, 2, 3, 4, 5, 6, 7, 8, 9, 10, 11

Scottish Curriculum Framework SKU 1, 2, 3, 4, 5, 6, 7, 8, 9, 10, 11, 12, 13, 14, 17, 18

Welsh Desirable Outcomes WKU 1, 2, 3, 4, 5, 6, 7, 8, 9, 10, 11

Northern Ireland Curricular Guidance NIKAE 1, 2, 3, 4, 5, 6, 7, 8, 9, 10, 11

Analysis:
Handling artefacts will encourage young children's critical thinking skills. You should concentrate on asking a number of questions about them.

❑ What is this?

❑ What do you think it was used for?

❑ What is it made from?

❑ Who would have used it?

❑ Where and when would it have been used?

❑ How does it work?

❑ What would do the same thing today?

❑ Is it old or new?

Artefacts can be hidden in feely bags or boxes, wrapped up as parcels with clues or one child can describe an object for others to guess what it is. Children can add or make labels to describe the objects, draw the objects, construct replicas of the objects from junk materials to use in their play or sort the artefacts into things that are the same or different. Comparisons can be made about size, shape, texture, condition, weight, colour or age.

Evaluation:
Ask children to match artefacts to photographs, pictures in books or stories. Set objects out in logical sequences to indicate age or technological development. A book such as Allan Ahlberg's *Peepo!* depicts home life in the 1940s. You could use this story with some artefacts to discover more about life in the past when children's grandparents were young. The book acts as a bridge from the children's own experience to that of new experiences of times past but it can also be used as a point of reference. Children could ask their own grandparents or family members about life in the past or an older person could be invited in to show children how particular objects were used. Children have a natural rapport and fascination with older people. Many older people love reliving their memories. Involving the two age

Assessment

❏ Note down the vocabulary children use when handling items from collections. (identification)

❏ What observations have they made? (identification)

❏ How did they sort objects and artefacts? (analysis)

❏ Have they begun to make connections between objects and artefacts? (evaluation)

❏ Have the children learned to handle objects and artefacts with care? (evaluation)

❏ What sorts of objects do the children collect? (evaluation)

groups in working together has enormous social as well as educational benefits.

Old or new?

Identification:

Ideally you need two similar objects as a basis for your activities. Which objects are old and which are new? How do they know? From two objects you can move on to three. Children will make their deductions based on their observations and on discussion.

Objects associated with household technology, such as irons, pots and pans, are particularly useful. Ideas about changes in technology such as lighting can be explored by looking at candles, oil lamps and light bulbs.

Artefacts associated with famous people - real or fictional - can be presented as evidence for children to work out who they belong to. Cinderella, for example, might have a duster, a clock set to midnight and an odd shoe. This kind of activity allows children to explore real objects with their senses, to talk and to listen and to use their imagination.

Analysis:

You may find that children go home and are motivated to find more objects associated with the past. The natural progression would be to set up a class museum and timeline. The collection can then be analysed to look at what has changed over time and why and how things work or worked.

Evaluation:

Through hands-on experiences such as this children will start to make more sense of the past. They will be acting as genuine historians - asking questions, examining evidence, sorting evidence to make sense of it, interpreting the evidence and using this to answer questions and to ask new ones.

Becoming an archaeologist

Archaeologists discover more about life in the past by exploring objects and sites that have been left in the ground. Children can go on a dig. What do they find? Turn your sandpit into an excavation site or hide replica objects in clay and allow them to harden. Children can then carefully dig using spatulas, trowels and paintbrushes to reveal objects from the past.

Using collections to develop a sense of place

Identification:

Collection tables where children can display objects, souvenirs or things they find are a useful asset in every early years setting. They can be linked to environmental or natural topics as well as to made objects. Encourage children to talk not only about what the objects are, but also where they come from. Find places on a large clear world map, a globe or a big atlas.

Analysis:

If you regularly change your collection table to reflect the curriculum you plan think about whether you can give it a geographical connection - foods from different parts of the world, different natural objects such as shells

and rocks from different places, coins from different countries. Do you make these displays static or interactive, involving the children in some way? Can children sort the objects into things associated with places near and far, the town or country, the beach or mountains?

Evaluation:

We all have personal connections with different places. Some children will have lived in different places, have visited them on holiday, they will wear clothes and shoes made in other parts of the world or have eaten food from another country. Some will also be aware of different currencies, writing and languages.

Children's toys and games are useful starting points. Toy vehicles can be linked to different landscapes. Which vehicles would you find where and why? How would you get from a to b? Where were these toy cars made?

Collections can be used to allow the youngest children to experience shape, texture and materials, to have indirect experiences of distant places and to make comparisons between places. Older children will be able to link their collections to maps, start to consider differences between cultures and link the objects in the collections to reference sources and information sources. Your youngest children are likely to collect information, older children will be involved in organising it and, later, analysing it.

Environmental collections

When children are outside exploring, they tend to collect objects of interest to take back inside. It is important to establish some ground rules about what and how children can collect, particularly if you are exploring the natural environment, to ensure that it isn't damaged as part of that process or to make sure children don't hurt themselves.

ACTIVITY

Character geography

Children readily identify with fictional characters they encounter in stories, videos and on children's television. They often have toys, videos, books and sometimes even clothes, bags and other memorabilia connected with a particular character. Children can become so engrossed with the characters and stories that they will listen to and watch them again and again and relive their adventures in their small world imaginative and role play. They find it hard to separate truth and fiction and live in a world that deals with both reality and make-believe.

Characters appear in every rhyme, story and song they learn and they often provide a useful way into developing children's geographical understanding, helping them to find out more about the world, within and beyond their experience.

Starting with a familiar character will enable children to act as geographical investigators. They will be able to

Curriculum guidelines

Early Learning Goals KU 1, 2, 3, 4, 5, 6, 7, 8, 9, 10, 11

Scottish Curriculum Framework SKU 1, 2, 3, 4, 5, 6, 7, 8, 9, 10, 11, 12, 13, 14, 17, 18

Welsh Desirable Outcomes WKU 1, 2, 3, 4, 5, 6, 7, 8, 9, 10, 11

Northern Ireland Curricular Guidance NIKAE 1, 2, 3, 4, 5, 6, 7, 8, 9, 10, 11

discover more about the world and how the characters and artefacts associated with them fit into particular locations. Some of these characters will live in a modern or reasonably familiar world. Some stories are set in places you can visit. Thomas the Tank Engine and the Island of Sodor, for example, is based in the Isle of Man. Others, such as fairy tales, will be linked to life in unfamiliar worlds, places that are far away in both distance and time.

Learning objectives

❏ To enable the children to find out more about imaginary people and the places and environments they are associated with;

❏ To enable them to discover more about people's roles and beliefs;

❏ To enable children to develop an appropriate vocabulary based on the characters explored.

Once upon a rhyme

Rhymes are an integral part of good early years practice. They provide excellent whole group starting points for sharing and discussion. They not only motivate children but are also full of information about people and places.

Children will learn to recite and enjoy a number of traditional nursery rhymes. All of these are set in particular locations. Take 'Jack and Jill'; they 'went up the hill'. Why do you have to go up and down hills?

What other directions can people travel in? What do they go to get? Where do they get the water from and why? 'This is the house that Jack built' helps children to explore the process of building. 'Dr Foster went to Gloucester in a shower of rain'. Where is Gloucester? What does a doctor do? What happened to him when it rained? The Grand Old Duke of York moves in different directions. 'Ride a cock horse' mentions Banbury Cross. Can you find Banbury on a map? 'Row, row, row your boat' - where to? What does downstream mean?

Similarly, more modern nursery rhymes, parodies on traditional rhymes and number rhymes can be used. Where did the five little ducks go to and why? Where did you buy the five current buns? Many anthologies, including *This Little Puffin* by Elizabeth Matterson, will be particularly useful for such work and rhymes linked to each topic should be collected and shared.

Back into fairyland

Identification:

Children encounter a rich repertoire of stories told and read in their early years setting. It will include fairy tales, legends and myths. Fairy tales are often moralistic, providing an exemplary behaviour code. They are exciting and rich in vocabulary. They can also be used to introduce children to geographical ideas such as location, mapwork and sequencing. Stories are woven together into adventures of characters set in geographical locations

within particular physical or economic circumstances. When such stories are shared the listener will start to visualise a landscape or mental image where the story is set which can be mapped, drawn or modelled in a 2-d frieze or a 3-d landscape model. These tales often contain journeys made across a landscape. In Snow White the story takes place in a number of locations (see map illustrated).

Analysis:
Certain tales are useful for exploring certain environments or concepts. 'Little Red Riding Hood' and 'Hansel and Gretel' explore woodland locations. 'The Hare and the Tortoise' involves a race and obstacles, which test the animals en route. 'Jack and the Beanstalk' is set in a castle above the clouds and can be linked to growing conditions, directions and the rooms in the giant's castle. 'Cinderella' can be linked to a sequence of events and places. 'The Enormous Turnip' and 'The Little Red Hen' can be linked to farming and the seasons and the tale of 'The Gingerbread Man' involves a number of people working within a landscape.

Evaluation:
The story of 'The Three Little Pigs' can be explored interactively in a Dorling Kindersley CD-Rom of the same name and is often used as a stimulus with three- to five-year-olds for work across all of the areas of learning.

Children's TV characters

Postman Pat lives in a rural environment. What do children associate with Greendale? Fields, hedges, animals? How does the weather affect him? What services does he provide for this rural community? Bob the Builder lives in a town, Thomas the Tank Engine on an island. Fireman Sam involves members of a fire brigade, people who help us in Pontypandy and Newtown. They wear special clothing, use particular equipment and rescue people in difficulty in a fictional Welsh community.

The Teletubbies live in Telletubby land inside a grass mound. How do they get in and out of their home? Does it have rooms? What is the weather and landscape like?

Rosie and Jim have all sorts of adventures aboard the Ragdoll, a canal barge and visit places to find out about different products and where they come from. Tilly, Tom and Tiny from *Tots TV* live in a cottage and explore their immediate environment but Tilly also speaks French.

Other stories

There is a huge range of excellent modern fiction set in a variety of places. It is likely that you already use stories as a starting point for much of your work. You may even have some big books which offer geographical learning opportunities. Books by Ruth Brown involve cats and are a rich source of information for environmentally based work.

Jez Alborough's *Where's my Teddy?* tells of a little boy hunting for his lost teddy in a woodland landscape which is dark and scary. Landscape, direction, feelings about places, routes, size and scale could all be explored through this story. Hunting and planting activities, giving directions and recreating the story would all extend children's knowledge and understanding of the world.

Mental mapping

Identification:
Even the youngest children have internalised their own ideas about what particular environments are like in the form of cognitive or mental maps. This is often shown through children's play, particularly their small world play with toys reminiscent of particular people, places and environments. Children can be asked to draw their own maps of journeys characters have made or places they live in. Board games can be produced so that children can locate particular features of stories by using simple finger matching, letter and number recognition techniques as the start of work on co-ordinates. Characters can be made into magnets

Mathematical Development

Distances
Size, shape and construction of homes

Communications, Language and Literacy

Rhyming words - huff and puff
Naming, listing objects in different houses
Different versions of the story
Sequencing events in story
Retelling story and giving instructions
Predicting what will happen next

Personal, Social and Emotional Development

Role play, drama
Recreating events
Feelings in role

Creative Development

Painting and modelling characters, homes and environments
Sounds, music making
Who's afraid of the big bad wolf?

The Three Little Pigs

Physical Development

Manipulative skills
Movement skills

Knowledge and Understanding of the World

Design, make and test three houses
Use programmable toy as wolf
Compare building materials
Explore similarities and differences
Design a symbol or picture for each house
Design a large floormap
Place three models on maps and move character models around
Give the wolf instructions to follow using directional and locational vocabulary

and moved around landscapes by using magnets underneath the base board and children can describe journeys made, providing excellent opportunities to assess their understanding of vocabulary and concepts.

Starting with a story

It is worthwhile breaking down a series of stories you regularly use about favourite characters in your planning documentation (see table).

Topic webs can be used to plan appropriate experiences (see page 47 - 'The Little Red Hen').

Visits and visitors can be included and processes looked at in terms of both the past and present. One character and his or her adventures can be linked to specific themes and vocabulary, providing a meaningful context for exploring many areas of learning.

Book _____ Author _____ Publisher _____					
Storyline	Skills	Concept	Context	Activity	ELGs

Creating a frieze about the story landscape with key words like farmhouse, mill, woods.

Mapping the story (farm - mill), describing the route the hen takes

What seeds are made into flour?

What other animals appear in the story? Sorting and matching animal pictures and sounds

Meeting a hen (life cycle of a hen)

How seeds grow - what is needed for growth?

Sequencing the events in the story

The Little Red Hen

Synopsis: Hen finds a seed and sows it. She tends the seedling as it grows, harvests it and takes the grain to the mill. With the flour, she makes some bread. None of the other animals help her but are keen to join in the eating of the freshly baked bread.

How do farmers grow seeds?

Making a 3d landscape of the story

How are seeds made into flour? Visit a mill.

How are crops harvested?

Eating bread you have made

How is bread made? Visit the baker's.

What is flour made into?

Finding out more about processes such as this.

Assessment

❏ Note down the vocabulary children use. (identification)

❏ What do the children know about imaginary people and the places and environments they are associated with? (analysis)

❏ Engage children in activities that enable them to construct a mental map of the locations and events in a rhyme, fairy tale or story. (evaluation)

Using a floor turtle to recreate journeys, describe position and direction

Resources

The Three Little Pigs Dorling Kindersley CD-Rom £19.99 (Dorling Kindersley Family Learning)

Catalogue: www.dk<None>.com

Jungle role play

Role play is a powerful tool for learning in early years settings. Role-play areas can be transformed into environments associated with past times or distant places which allow your children to develop their knowledge and understanding of the world by travelling in time or to distant locations in a way that is practical and meaningful for them.

Starting a theme by creating a role-play environment will enable children to act as geographical investigators and historical detectives. They will be able to discover more about the world and how the characters and artefacts associated with it fit into particular locations and times.

Places or museums that children have visited or scenes from photographs can be recreated in your setting. Always involve the children, whether it be making Victorian homes or castles with card, paper and paint or replica artefacts and costumes. Make and explore environmental habitats in your locality or further afield - underwater worlds, woodlands, deserts, grasslands, Arctic or Antarctic environments, beaches, mountainous landscapes or tropical rainforests. Careful planning is needed to create a well resourced and attractive role-play area that supports your theme, your learning objectives and at the same time offers a wide range of play rich opportunities.

Learning objectives

❑ To enable children to discover more about the world, past and present, through play;

❑ To use the play context to allow children to recreate past times and distant environments;

❑ To encourage the children to appreciate different times and cultures;

❑ To enable children to develop an appropriate time and place vocabulary through their play.

Rumble in the jungle

You will need to provide:

Animals masks (you can make these in a previous creative session)

Old fake fur coats, rugs or fabric

Camouflage clothes

Safari helmets, clothes, cameras, binoculars, water bottles, rucksacks

Curriculum guidelines

Early Learning Goals KU 1, 2, 3, 4, 5, 6, 7, 8, 9, 10, 11

Scottish Curriculum Framework SKU 1, 2, 3, 4, 5, 6, 7, 8, 9, 10, 11, 12, 13, 14, 15, 16, 17, 18

Welsh Desirable Outcomes WKU 1, 2, 3, 4, 5, 6, 7, 8, 9, 10, 11

Northern Ireland Curricular Guidance NIKAE 1, 2, 3, 4, 5, 6, 7, 8, 10, 11

First aid kits, insect nets

Butterfly nets

Notepads

Pictures and sounds of jungle environments

Camouflage net

Toy snakes, spiders, crocodiles, and so on

Picnic and camping equipment

Dinghy, canoe, oars

Maps, compass, radio, computer

Songs, stories, books and poems

Non-fiction posters, videos and books about rainforest environments

Involve the children in creating this role-play area. Use pictures to spark off ideas about what should be there and to plan, design and make the environment suitable for your setting. This will not only develop children's knowledge but also their vocabulary.

Once established, props can be placed in the area to echo a particular learning objective and encourage children to explore and live in the environment they have created.

For those who find it difficult to engage in role this theme could be explored through a toy, such as Safari Sam or Hiking Heidi (available from Boots the Chemist).

Children can explore new vocabulary and the sounds associated with this environment. They can work in groups to make decisions about routes to follow, where to set up camp and how to care for the tropical environment and wildlife associated with it. Work on animals can be linked to colour, silhouettes, size and number. Magnifiers, binoculars and telescopes can be used to make things bigger and smaller. Children can manipulate tools and equipment, move like animals, climb, row, step across stepping stones, swim across crocodile infested rivers, wade through swamps or swing from tree to tree. They can devise routes and trails and provide maps for others to follow.

Finding out more about the rainforest

Evaluation:

Television documentaries and wildlife programmes, commercial videos produced by National Geographic, easy-to-access picture non-fiction texts, CD-Roms and the Internet can all be used with adults to find out more about the rainforest. The World Wide Fund for Nature (Panda House, Weyside Park, Godalming, Surrey GU7 1XR) provides resources, toys and ideas linked to the preservation of wildlife and natural habitats. This includes small board books on topics such as the panda and soft fabric or rag books shaped like animals in danger of extinction which are fun to look at and share. They contain a strong environmental message.

Globes and atlases are appropriate for the youngest children to feel and explore. They will look at them for recognisable colours, shapes and images. Which animals live in particular environments? Can we find the answers from these sources? What is the environment like there? What is the weather like? Where do they live? How do they feed? How do they bring up their young? Who are their

predators? How do they hide from danger?

You can introduce simple work on habitats, food chains and life cycles. Children can become different animals who live in various parts of the world. They can be transformed with face paints, masks and costumes. They can physically arrange themselves into food chains or this can be shown on displays or with 3-d hangings. Talk about what animals need to live (water and food) as well as which animals eat other animals and which eat vegetation.

Children can listen to soundtracks of animal noises and guess the animal or match it to pictures. They can make the relevant noises and learn to move like particular animals.

Play animals can be sorted to match environments. Concepts such as camouflage can be explored when thinking about how animals adapt to their landscape. Hide and seek games to find camouflaged animals hidden in pictures will help explore concepts

related to the colour and shapes in patterns on animals.

Different environments could be explored in similar ways through stories such as Helen Cowcher's *Antarctica* or the adventures of Penguin Pete or Pingu. Children could explore which story they felt was most realistic.

Green issues and the earth's sustainability can be explored on a local scale, studying minibeasts and local habitats in your immediate surroundings to exploit the scientific aspects of Knowledge and Understanding of the World.

Assessment

❑ Note down the vocabulary children use. (identification)

❑ Observe the children as they play. What do they use and say? (analysis)

❑ What misunderstandings do they have about life in the past or different places? (evaluation)

The circle of life

All early years settings cover work on the seasons at some time. It is an obvious theme to use when developing children's knowledge and understanding of the world as it impacts so clearly on both our lives and the environment around us.

The seasons are distinctive periods of the year in the UK, characterised by specific features such as the length of the days, changing the clocks, the weather we experience and what we wear and do. Seasons will be different in other parts of the world.

We tend to associate particular events and activities with particular times of the year - holidays in the summer and festivals such as Christmas in winter. For other cultures, festivals such as Eid can vary in their date but they, too, link into natural rhythms associated with the moon.

Seasonal change can be detected through a study of the environment. Natural events such as the first signs of growth in spring or the changing of leaf colours and the loss of leaves from trees in autumn are easily detectable and something children enjoy recognising and asking questions about.

Your children may have strong feelings about the seasons and will already associate particular events that happen around them or things they can or can't do with them.

Learning objectives

❑ To recognise the features associated with particular seasons of the year;

❑ To be aware of seasonal change in the environment;

❑ To know that particular weather conditions are associated with particular seasons;

❑ To know that regular events in their lives and the lives of others link to the four seasons;

❑ To develop a seasonal vocabulary.

The seasons of the year

Identification:
In the UK, we experience four distinct seasons - spring, summer, autumn and winter - throughout the year, even though it sometimes seems as though we experience them all in one day! Children need to learn the names and sequence involved so it is important for you to identify the cyclical nature of the process, explain which season follows which and help children to recognise the signs associated with the arrival of the next season. How do trees change in spring, summer, autumn, and winter? Which months of the year go with which season? Which season is their birthday in? When is Bonfire Night?

This can be depicted visually through a simple seasonal clock. Divide a circle into four sections of colour associated with the different seasons, for example green for spring growth, yellow for the summer sun, orange for the autumn leaves, and red for winter festivities. Make a single hand from card which you can move from season to season. Add pictures to illustrate the activities

children take part in, the weather conditions, what they wear, children's birthdays, national events, objects or pictures of natural features associated with particular seasons.

The life-cycles of pets, other animals and features of the environment are also connected to the seasons. Soil and water outside are affected by the weather - puddles will freeze over and ice will melt. The soil will be warm enough for certain things to grow in or be too wet to dig. The children will hear birds singing in the morning in spring and see them fly away to warmer climes in the autumn. They will associate some birds, insects and animals with particular seasons of the year - chicks and lambs with the spring and hedgehogs and squirrels with the autumn. All these changes can and should be explored through the senses outside, but you can also explore the seasons through secondary experiences. Particularly useful are:

❑ **Charts, pictures and books** with seasonal images of the environment, season by season.

❑ **Stories**

The House on the Hill by Phillipe Dupasquier (Penguin). This book consists of a double-page spread and a single word - the month of the year. The illustrations are full of detail about the weather and seasonal changes and the effects these have on family life. The family lives in a house on the hill, in an isolated rural area. Children can compare their experiences with those of the fictional family in the story.

Curriculum guidelines

Early Learning Goals KU 1, 2, 3, 8, 9, 10, 11

Scottish Curriculum Framework SKU 1, 2, 4, 5, 6, 8, 9, 10, 11, 13, 14

Welsh Desirable Outcomes WKU 2, 3, 4, 5, 6, 8, 9, 10, 11

Northern Ireland Curricular Guidance NIKAE 2, 3, 4, 5, 6, 7, 12

This theme appears as a distinct statement within the Welsh guidance. Children should: 'have a basic knowledge of the seasons and their features'.

In contrast, *The Year in the City* by Kathy Henderson (Walker Books) celebrates seasonal change and the hustle and bustle of life in the city through the months of the year. Children will identify with supermarkets full of chicks at Easter and illuminated decorations in the shops at Christmas. The text is designed to be read aloud and could form a monthly focus for work in your setting.

The Shepherd Boy by Kim Lewis (Walker Books). This book explores the life of James, the son of a shepherd, who wants to become a shepherd like his Dad but is told he isn't old enough. The story follows James through the passage of a year in his life on the farm focusing on the different seasons of the year.

❏ Songs, rhymes, sayings

Much of the repertoire in early years settings links to the weather and seasons. Check through the songs and rhymes you use. Which of them link to the seasons and how they affect us?

❏ In the news

Video clips or newspaper pictures which show the effects of seasonal change on our lives can be shared - roads blocked by snow in the winter, farmers' crops destroyed by drought and water hose bans in the summer.

❏ Weather charts

Most settings keep weather charts which feature words and pictures associated with times of the year, days of the week and the weather conditions. Symbols are used to represent different types of weather and children will have seen similar or different ones used on weather reports on the television.

❏ Popular TV programmes

The weather and seasons form an essential backdrop to the life of fictional characters children watch or read about regularly - Percy the Park-keeper and the animals in the park, Bob the Builder. Postman Pat's adventures take place on windy, foggy, sunny and snowy days. Characters in Jane Hissey's *Old Bear* stories also have adventures in different seasons such as *Jolly Snow*.

Nature trails

Identification:

Each season presents an opportunity to take a walk around your grounds, observe changes taking place in your garden area or explore a range of natural and wildlife habitats in the immediate locality or further afield, perhaps through a visit to a park, a farm or a garden centre.

A nature trail can be marked out in your outdoor area using arrows, string to follow or numbered locations. Link it to directional, locational and sensory vocabulary for children to follow such as walk, look, touch, through, around, up or down. Make some simple cards depicting actions the children need to take at particular locations, for example cards with arrows pointing in different directions or with eyes for looking, ears for listening, and so on. Turn them over to reveal the action at relevant points on your walk. The youngest children can be led around the trail, whereas older or more able children will enjoy the idea of a treasure hunt and trying to find particular things in the environment.

Children can be blindfolded and led by an adult. Mark stops with an opportunity to discover more about the seasonal environment through their senses - touching textures, smelling herbs and flowers, walking on different surfaces and listening to the sounds made. Such activities will enhance their awareness of the natural environment, help them to communicate their views about it and hopefully learn to appreciate and care for the environment and the living things within it.

Sensing activities can be recorded with tape or video recorders, cameras or digital cameras. You can make sketches and wax crayon rubbings of textured objects. Carry out a survey by placing a PE hoop on the ground and asking children to look closely at that particular part of the environment.

The things children notice and record can be displayed on tables or friezes when you return inside, or made into books, seasonal diaries and photo albums.

A scavenger hunt

Analysis:
Children are natural collectors. They love to collect objects they find. Give them containers and ask them to collect objects associated with particular places and seasons. Ask them to find a number of objects or objects of a particular shape or texture. These can then be used for groups of children to sort and classify, grouping objects into, for example, natural and made, matching objects with similar patterns, counting objects with the same shapes or colours or measuring and comparing their findings.

Young children can complete observational drawings, rub the surfaces and shapes or make collages of the objects they find in a particular place. They can print with objects they find or weave grasses and leaves together.

Display photographs or posters of particular seasons and place appropriate objects nearby on a nature table. The accuracy of children's observational skills and vocabulary will develop as they engage in more of

Our sensory walk

What did we hear?

What did we see?

What did we feel?

What did we smell?

We liked/disliked this environment because...

We can care for this environment through...

these sorts of activities as the year progresses. The objects they find can be placed in feely bags or boxes and children can use their sense of touch to describe them for others to guess what they are. Encourage them to use words such as hard and soft or rough and smooth.

Try to take photographs of the children's explorations to remind them of what they found where. You can add them to a basic plan of the route taken to show where objects were found.

When children are involved in exploring the environment you will need to explain to them that some things are safe to collect but others are not. Remind them constantly about health and safety issues and ensure hygiene routines are followed.

Evaluation:
Children will link what they find to their existing knowledge or may be motivated to find out more through the use of books or adults as sources of information. Learning *through* the environment will allow children to further their knowledge *of* the environment.

There are also important environmental issues to consider. Children need to learn to care for their environment and understand that they mustn't collect things that would cause environmental damage. They should also be encouraged to have an opinion about their environment and start to notice things that they like or dislike about it, such as litter.

Helping children to develop an awareness of living things is an important dimension in this particular area of learning. You can try to make them more aware of the earth's natural processes and cycles. Make the most of spontaneous opportunities as well as planned activities to nurture an appreciation of natural beauty and a sense of awe and wonder about the world in which they live.

How the weather affects our lives

Identification:
Everything we do is linked to the seasons and to the weather. Particular types of weather are associated with particular seasons. Fog, wind, snow, ice, rain and sun can all be experienced through first-hand sensory experiences. Natural connections can be made which relate directly to the children's lives. Will they need their wellies or coats on? Why or why not? Children can also follow the lives of birds and animals in each season as they are born, change and hibernate or migrate, for example.

It is important to encourage the increasingly accurate use of appropriate weather and seasonal vocabulary, to carry out simple experiments and make recordings. Children can learn weather rhymes and sayings and use these as prompts to discuss different kinds of weather and seasons.

Analysis:

Most children go on holiday in the summer. Which clothes would they pack in their suitcase and why? Where are they going on holiday? Will it be hotter or cooler? What will they need?

Can they dress teddy for different seasons or can clothes in the role-play area be matched to the seasons of the year? Does your role-play area have a window so that children can see outside? If not, place a painting of a different season up within it so that it looks like a window. Remember to change the picture regularly to depict a variety of weather types. Hang up a calendar showing seasonal pictures.

Evaluation:

Encourage children to consider weather conditions as part of their role play, for example if they are taking the baby doll for a walk in the pram, what will she need to be dressed in? Will they need to take their umbrella with them?

Assessment

❏ Do children use the correct seasonal vocabulary? Note down which words they use. (identification)

❏ Can children pin the correct picture to a window in a play corner, find the correct pictures for the weather chart, find the correct event image, for example Father Christmas for December, to depict the seasonal changes? (identification)

❏ Can children match pictures of events to the seasons? (analysis)

❏ Can they dress a teddy for different seasons or pack a suitcase or schoolbag with things they would wear in different seasons or appropriate equipment such as swimming gear? (evaluation)

❏ Do children associate changes in the environment or events with particular times of year, for instance sweeping up leaves in the autumn or watching tadpoles turn into frogs in the spring? (evaluation)

From raindrops to rivers

Water is an essential part of our lives. We drink it, swim in it and sometimes even need protection from it. Our homes are designed to shelter us from rain and floods; lifeboat crews work to rescue people from storms at sea. There are many traditional stories associated with water in all its forms. You can share them with young children to explore life now and in the past and learn how events connected with water have shaped people's lives then and now.

Water play is a regular feature in most early years settings. Children enjoy playing with water and through play can learn much about its unique properties. Without water we cannot survive - but do the children know that? A water theme linked to Knowledge and Understanding of the World could be explored through a number of focused questions which will not only plan and structure the theme but will also allow you to assess what the children know and understand. Adults can work with children through guided exploration of the local environment or through the use of materials readily available in any early years setting.

Learning objectives

❏ Recognise the properties associated with water;

❏ Be aware of water in the environment;

❏ Know that regular events in their lives and the lives of others link to water;

Curriculum guidelines

Early Learning Goals KU 1, 2, 3, 4, 7, 8, 9, 10, 11

Scottish Curriculum Framework SKU 1, 2, 3, 4, 5, 7, 8, 9, 10, 11, 12, 13, 14

Welsh Desirable Outcomes WKU 1, 2, 4, 5, 8, 9, 10

Northern Ireland Curricular Guidance NIKAE 1, 2, 3, 4, 5, 6, 7, 10, 11, 12

❏ Develop a water word vocabulary linked to the work you plan and undertake;

❏ Know that people and animals need water to sustain life;

❏ Know where our water comes from and that we should use it with care.

What do we use water for?

Identification:

Through discussion, encourage the children to think of the different ways they or their families use water - for washing, drinking, cleaning, cooking, playing, travelling on, and so on. Find pictures depicting different types of water use for children to cut out or sort. This activity can also be linked to locations and times of the day. Your regional water supply company may have an education department which can supply information and resources. Keep a water diary - how many times have we used water this week and why? Involve the children in tasks that use water - diluting the orange juice at snack time, dissolving jelly for a party, making ice cubes or ice lollies when it

is hot and washing their hands before and after certain activities.

What do other people use water for?

Analysis:

How is water used in places other than your setting - where parents work, on farms, in factories, or in different parts of the world? Use visual images to find out what the children already know. How do they know? What do they think? Do you have a good collection of pictures of water?

Children can use small world equipment to enact how firefighters use water to put out fire. Try to arrange a visit to a fire station or see if the local fire service can bring an appliance to your setting. Encourage children to think about where the firefighters get their water from and what might happen if there wasn't enough water available. Can they find a local hydrant sign? This could then be linked to the role-play area by supplying hoses, toy fire extinguishers, buckets and perhaps a cardboard box fire-engine so that the children can put to use the knowledge they have acquired. Do the firefighters need protection from all the water they use? What do they wear?

Water can be harnessed to help us, for instance, to power waterwheels or mills which grind corn, but it can also create barriers in the form of rivers and seas which have to be crossed by bridges or ferries.

Some water trays have canal and lock systems which allow children to

Assessment

❏ Can children identify when and why they have used water, for instance to wash their brush, clean their hands? (identification)

❏ Make a note of water words children use to check their developing vocabulary. (identification)

❏ Can the children recognise different properties associated with water, such as ice, waves, rain? (analysis)

❏ Can they name different features of water in the environment, such as river, pond, stream, for example pointing to a poster and identifying the water associated features? (analysis)

❏ Can children recognise that pets such as fish need their water changing regularly or that pets and plants would die if we left them without water? (evaluation)

explore how water is used to transport boats, people and cargo. This was an important aspect of life in the past. If there is a canal in your locality, arrange to go for a walk along the towpath and hunt for clues of the past or watch how a lock actually works. (Enlist extra help from parents and volunteers to ensure a high adult: child ratio.)

If you have an outdoor water supply, children can use this in their free play. Does your setting save water in water butts? Do children have access to it? If so, they could collect it in buckets to mix with sand to make mortar for a project on buildings. Can they use a watering can to collect water and water the flowers in the garden?

Use stories to show children what water can do. Phyllis Krasilovski's *The Cow Who Fell in the Canal* is set in the Netherlands and depicts a flat landscape with windmills and canals. Stories can also be a useful way of transporting children to different environments where water is more scarce, for instance V Aardem's *Bringing the Rain to Kapiti Plain*, or looking at environments where water is

more plentiful, as in Helen Cowcher's *Rainforest* and *Antarctica*.

Water has specific cultural and religious significance to certain peoples. Christians, for example, regard the River Jordan as a special place. Children can also get involved in helping people in parts of the world where water is scarce or where flooding has had disastrous consequences by raising money in response to appeals.

Children may see evidence of pollution in their locality or will see and hear news stories about environmental issues. This provides a good introduction to discussion about how dirty water can harm animals, plants and us as well as what happens when there is too much or too little water.

How does water move?

Evaluation:
Children will spend time playing with water in your setting, but do they know how it moves in the environment? They love to go outside and pour water from containers, watching where the water goes and what happens to it. Do they notice that any water they spill eventually dries up? Or that if they pour too much water in one place it becomes water-logged? That water tends to follow the direction of a slope and move downhill? That water reacts differently when it encounters different surfaces?
If there is a stream with a bridge nearby, take children on a walk there to play pooh sticks. If not, read them the story or watch the video of *Old Bear's Boat Race* (Jane Hisssey).

Where does our water come from?

We are fortunate that clean water comes directly to us from our taps. Do children know how it gets there and how this links to the water cycle and the journey of a raindrop? This wasn't

always the case and isn't true for all parts of the world. Do children know where water came from in the past? Can they guess? Do they know and use words like river, stream, pond, sea? Sing 'Jack and Jill' and talk about why they went up the hill to fetch a pail of water. Where did they get it from? If any children have visited historic buildings, for example a castle, what did they notice about water and how it was acquired by the people who lived there?

Children often know lots of isolated pieces of information about water but you can help them bring this information together in a logical way.

Rainwater can be collected and measured in terms of full and empty containers or of more and less rainfall collected in different places. From your discussion with children, produce a long strip of card. At one end, write 'rain' as the start of the water cycle; carry on adding the words children know and use - clouds, river, sea - in a logical order and ask the children to illustrate them. Add information they have discovered from their explorations, for example about how water gets into the tap. Tape each end of the card strip together to form a circle and illustrate the continuous nature of the water cycle.

The journey of a raindrop can also be followed in the sand tray. Make a landscape with hills and valleys, adding pebbles and toys to make it more realistic. What happens if there is a sudden downpour on the hill?

Where does dirty water go?

What happens when we pull the plug out of the sink? What do children think happens to the water? What happens to the rainwater that falls on the roof? Children can observe this at first hand and explore outside to discover evidence of drains, manholes, pipes and gutters.

What Learning Looks Like... 55

All about me

Memories are made of this

Young children are egocentric. They are the centre of their own worlds and are fascinated by themselves and the people and places around them. They are curious about where they came from, how they have grown and whether adults did the same things as they do when they were little.

There are obvious connections within such a theme between a growing knowledge and understanding of the world and a child's personal, social and emotional development as well as their physical development. Children tend to notice the differences rather than the similarities between people's lives. An in-depth study of themselves will allow them to think about other people as well as themselves and to find out more about past and present events in their own lives, and in those of their families and other people they know. Through such a topic, children should also begin to understand more about their culture and beliefs and those of other people.

Learning objectives

❏ Identify similarities and differences between themselves and other people;

❏ Understand that people grow and change;

❏ Be aware of how they relate to other people and places;

❏ Develop an 'all about me' word vocabulary, linked to your children, your setting and the activities selected.

My personal passport

Identification:

It is essential that all adults working with children get to know them and their likes and dislikes. Knowing and remembering names and information about people is a part of everyday life. Take

photographs of your children, write out their names and let them match the names to the faces. How many children have the same name? How do you tell them apart? Can they describe each other? Recognise each other from descriptions? Play 'Who am I?' Activities can involve recording facts about themselves - height, weight, hair and eye colour. This can lead on to finding out more about their names, addresses and where they were born in order to spot similarities and differences. Make passports to use in role play at the travel agents, at a holiday destination, or an airport or ferry terminal. Some computer programs start with an identification page which

---

Curriculum guidelines

Early Learning Goals KU 1, 2, 3, 4, 8, 10

Scottish Curriculum Framework SKU 1, 2, 4, 5, 6, 7, 8, 9, 10, 12, 14, 18

Welsh Desirable Outcomes WKU 1, 4, 5, 8, 10, 11

Northern Ireland Curricular Guidance NIKAE 2, 3, 4, 5, 6, 7, 9, 10, 11

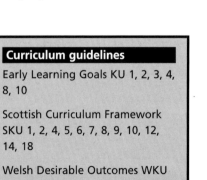

Callum's Christening

Callum's baby brother.

Callum's 4th birthday.

Callum's 1st holiday.

requires children to log on with their names and identify someone they look like. The Dorling Kindersley CD-Rom, *All about me,* is useful for children who can recognise letters and words or could spark off a few ideas for adults working on this theme.

It is important for children to learn to recognise names. Names are important in building up relationships with children and between the groups of children in your setting. What names are they known by? Does anyone in their family have the same name? Which is their family name or surname? Children need to see their name written on a regular basis. Think about introducing activities to promote name recognition as part of your everyday routine. For example, as children arrive at your setting, you might ask them to find and move their name on a task board to show the activity they would like to do or whether or not they are staying for lunch that day.

Vocabulary acquisition is a major feature of this theme and should be recorded to assess children's levels of development. Children will need to use language accurately in order to explain what they start to understand about the process of change.

Early memories

Analysis:

Children love to look back at pictures of themselves when they were younger to see how they have changed. Most families have photos of memorable events - birth, christening, first holiday, first bike, birthdays, and so on. Ask children to bring these in to sequence their own personal timelines and those of other members of their families.

Children need to understand that photographs can be used to record how things change. They are a useful way of providing memory joggers about past events such as holidays, how we celebrated Christmas, the

millennium or particular birthdays. They can also be used to see how children and their families have changed. What did they used to look like? What could they do? What were their favourite toys? Children should be encouraged to talk about how they have changed.

You can carry out a similar exercise with family photo albums, comparing photographs of parents and grandparents. Some of these could be in black and white; very old pictures will be sepia. Most children will assume that black and white photographs are old, colour ones more recent. Explain that you can still buy and use black and white films (though you may find that the processing costs more). It might reinforce this message if you can take some of your own. Teach children to look carefully at the details in the photographs to find the clues to life in the past.

Happy birthday!

Most children will celebrate a birthday during their time in your setting. The years mark a child's growth and development and children are proud of becoming older. Celebrations are held and candles used to commemorate the number of years of life. Children can work out who is the oldest and the youngest and share how their birthdays were celebrated at home. Not all cultures celebrate birthdays in the same way and care needs to be taken to value these rites of passage with different children.

Annual events

During the year, events in our national cultures will be celebrated which commemorate particular occasions. Children will be keen to discover more about these events and to experience the celebrations that go with them so try to include them in your forward planning.

The generation game

Grandparents provide a ready source of information about how things have changed and what life was like in the past. They often have the time to talk and to share experiences and photographs with young children. They will also introduce children to vocabulary such as 'When I was little', 'That was when we got a new car', or 'That was an old building'. If few grandparents live locally an older member of the community can be adopted to act in this role. It is useful if you can provide pictures, artefacts or photographs in sequences of three for grouping according to the generations - themselves, their parents and their grandparents.

Precious things

Throughout our lives, we hold on to memories and treasures of particular significance to us. Children need to learn to handle such items that are precious to others with care. Many young children will have a favourite toy or a comfort blanket that they are still inseparable from. Adults can share their treasures and discuss when and how they acquired particular objects and what they mean to them. Bob Graham's *Red Woollen Blanket* introduces us to Julia. As Julia grows from birth to starting school, her blanket shrinks as numerous things happen to it.

Parents often keep memorabilia of a child's past and children will be amazed to look at clothes they wore, their first pair of tiny shoes, their christening robe or first birthday cards. They will see how they have changed and will be able to sequence and explore things that are precious to them now in comparison with then. They will be learning to handle and interpret information from evidence related to themselves. They will also start to develop concepts associated with old and new, now and then.

A day in the life

Photographs of a day in the life of children in your setting can be used for sequencing using a washing line and pegs. Use them to stimulate a time vocabulary, such as yesterday, dinner, afternoon, today, soon, next, and to work out what happens before or after snack time. It is these events in a day which give children a sense of order and time. Days will be ordered for children by events such as getting up, eating breakfast, bath time and bedtime. Objects associated with these events can be sequenced and grouped.

Specific events can be timed and you can make connections with clocks - particularly the o'clock. When baking, time how long the cakes need to be in the oven. You could time how long it takes children to tidy up. Clocks can become part of your role-play area. Introduce children to analogue and digital clocks. It is useful to have a 'teaching clock' on which children can move the hands around to see what happens. Young children won't be able to tell the time but will know something about numbers, hours and minutes. This will link into other vocabulary they will learn about the days of the week, months of the year, the seasons and words such as tomorrow, yesterday, past, present and future.

A day in the life of a famous person in the past can also be explored through story-telling or reading. Most stories are set in the past and, whether based on real or imaginary life, will allow children to access information about life in the past. Events in someone's life can be acted out.

History around us

Evaluation:
History is all around us. Walks through a village or town enable children to play at being historical detectives -

looking for clues to life in the past, datestones and buildings which the children think are old. New developments such as an extension, new window frames or the construction of a new supermarket change places and it becomes difficult for us all to remember what places used to be like. Children's understanding of the passage of time is rather different to our own and they will need to be introduced to many and varied experiences to enable them to build up their knowledge of life in the past. Many museums now cater for young children and actors or interpreters will explain what life was like and introduce sensory experiences. Visits can be recreated back in your setting through role play. Castles or Victorian homes can be built from junk materials and children can learn about the past through play. They will enjoy dressing up, handling artefacts, recreating daily routines of people from the past and spotting similarities and differences between their lives and those of people in the past.

This clues approach to life in the past could be explored through burying a time capsule in your grounds. Ask the children to help you compile an accurate record of what they are like and can do at the start of the year. This can be dug up at the end of the year to see how they have changed.

Play at being archaeologists by hiding a number of objects in the soil outside or in the sandpit. Children can excavate them carefully with trowels and brushes. What have they found? What do they think they are and what have they been used for?

Other people's histories

It may be possible to explore the histories of people they know through photographs and discussion. You could compile evidence bags based on familiar people. Open each bag to find clues as to who they might belong to. Children who need support can be helped through appropriate questioning.

We depend upon surviving evidence from the past to tell us about people who lived longer ago. Resources can be borrowed from museums or collected at car boot sales. Use them for 'Guess what?' and antiques roadshow type games and as props in role-play areas based around life in the past. Guesses can be confirmed by comparing objects with people's memories or with secondary evidence in books and pictures. Much of what survives links to items that are no longer used in domestic life, often because of changes in technology. Young children can handle old implements associated with cooking, cleaning and washing as part of an investigation of household technology now and then. Changes in domestic washing techniques can be explored through Allan Ahlberg's *Mrs Lather's Laundry*. Life in a 1940s home can be explored inside and out through Janet and Allan Ahlberg's *Peepo!* Events in these stories can also be sequenced to help develop children's ordering skills.

Stories provide an essential shared experience which encourages listening and questioning. They can provide an effective way into more distant times and places and enable children to compare the lives of others with their own lives, communicating knowledge in an easy-to-understand format. Questions can be asked about stories shared with children - When do you think the story happened? Why do you think it happened a long time ago? Are there any clues in the pictures? Is the story true or make-believe? Certain stories are useful for developing different aspects of children's historical understanding.

Allan Ahlberg's *Starting School* (Picture Puffin) can be used to explore time, sequence and chronology. John Burningham's *The Baby* (Cape) can be used to look at change in a pupil's own life. Marie Williams' *When I was Little* (Walker Books) can be used to explore a Granny's recollections of her childhood. Traditional stories such as 'Little Red Riding Hood' or 'Jack and the Beanstalk' can be used to explore life in the past for other children.

Tomorrow's clues today

Evaluation:

Young children find it difficult to comprehend that other children's lives are different from their own, let alone how different life must have been in the past. It is important to emphasise that everyone has the same basic needs and is likely to have the same sorts of memories - where they live, favourite foods, activities, people and games. We can find out about children's lives in the past from the objects they leave behind - buildings, paintings, pieces of writing, tools and toys. With the children in your setting, put together a box containing things which are typical in terms of their memories and favourite things today. What clues would they provide for others about this particular group of children and how we live today?

Children can compile a memory scrapbook of activities and work undertaken with you that they can be presented with when they leave school. This provides a wonderful memory source to look back on with parents, carers and with future teachers.

Assessment

❑ Can children communicate information about themselves? (identification)

❑ Can they spot similarities and differences between themselves and others? (identification)

❑ Can they base their views on evidence from artefacts, pictures, and so on? (analysis)

❑ Can they sequence three items connected with themselves, their parents and their grandparents? (analysis)

❑ Can they explain how they have changed? (evaluation)

People who help us

The world of work

Young children enjoy dressing up and pretending to be other people. They encounter many different adults in their daily lives. Characters they see on children's television have specific roles - Bob the Builder, Fireman Sam, Postman Pat - and they all help everyone they meet. You can make the most of all these opportunities to develop children's knowledge and understanding of the world of work.

Learning objectives

❑ To identify people who help us;

❑ To understand what the people who help us do, use and wear;

❑ To be aware of how they relate to other people and the roles they have;

❑ To develop a 'people who help us' vocabulary, relevant to the particular choice of theme.

Learning through play

Children's toys recreate different places and occupations. As they play with Duplo, Lego, Playmobil, toy garages, farms, houses, road layouts, train sets, airports and playmats, they will be exploring and developing their knowledge and understanding of the world. Adults will need to observe and support the play by encouraging children to talk about the ideas they have about people and how they help us.

Who works in our community?

Identification:
Each early years setting will have its own community of children and adults. Who is who and what do they do? Together you could produce a

| Curriculum guidelines |
Early Learning Goals KU 8

Scottish Curriculum Framework SKU 7

Welsh Desirable Outcomes WKU 5

Northern Ireland Curricular Guidance NIKAE 8, 9, 10, 11

photoboard of people and roles for your setting. Where do these people work? Does your setting have offices, classrooms, kitchens, and so on where different people are based? Children can illustrate the tasks people carry out with their own models, drawings and writing, using ICT where appropriate. Could some of this work be included in parent handbooks you provide to enquirers and new parents?

Analysis:
Are there people who help you get to your setting safely? What does a lollipop person do? Encourage children to act out this role in your outdoor play area by drawing a road system, adding pavements, road signs and zebra crossings. The children can use large wheeled toys to be the vehicles and pedestrians can push prams. What are the difficulties? How are we helped? What do we do? This role could also be explored through the use of small world play and tabletop equipment such as Playmobil or with floor mats and play people. When exploring road safety issues link them to the children's own world and routines and the importance of keeping safe.

Visits can be made in the local community or visitors can be invited in to talk about their role.

Key roles

The key roles to explore are:

the shopkeeper

the dentist

the farmer

the teacher

the postal worker

the factory worker

the mechanic

the refuse collector

the doctor

the nurse

the cook

the hairdresser

Props and resources:

Tables, chairs, cloths, flowers, menus, serviettes, candles, crockery, glasses, cutlery, food, tea and coffee pots, aprons, hats, tea towels, notepads and pencils, a serving hatch, wine bottles, sugar bowls, umbrellas, posters, newspapers and magazines, a till, money, credit cards and cheques, condiments, plants, music.

The best way to explore these workers and the environments they work in is through the first-hand experience of a visit followed by an opportunity to explore what they have learned through role play. Photographs and videos can be made of visits and these can be used to help convert the existing equipment and space you have into an exciting environment. A shop can be turned into a supermarket with packets, bags, cash tills, shelves and money. Children can take on different roles and adults can structure their play to ensure high quality learning, and genuine development of knowledge and understanding of the world occurs. The supermarket can reflect different aspects, such as a bakery one week, a fruit and vegetable area the next. The children can look at who delivers what to the supermarket, where things come from and how they get there as well as what people buy and how they carry it home.

If a new supermarket is being built nearby, children could watch this process and think about the building, the car park, and why certain sections of the supermarket are placed together. This play dimension needs to be carefully planned to maximise learning opportunities across the curriculum and link to themes and other experiences.

The doctor, dentist and nurse help us in different ways and children will have had regular experience of the baby clinic, the health visitor and perhaps

even the hospital. The doctor's or the hospital is a popular focus for unstructured play. This play can be enhanced through visits and by recreating experiences linked to their whole development - allowing the children to take on and act out different roles. A baby clinic with its weighing and measuring equipment and eye charts would link to topics on growth or 'When we were babies' and link to the 'All about me' theme. A good way of undertaking effective assessment is to place a tape recorder in the role-play area to record children's conversations.

Pictures, play equipment, costumes, hats and props will be required to match your chosen topics. The Dorling Kindersley book *Jobs People Do* would be a useful asset. Each A4 page identifies a different job people do and the clear simple text explains exactly their role. The pages are illustrated with photographs of young children dressed in role and attention has been given to ensure the book presents no gender stereotypes.

Links across the curriculum

Evaluation:
The strength of well-planned and well-resourced role play is its potential to span all areas of learning. Through setting up a role-play cafe, for example, children will not only learn about the place and the people who work there but also explore the different tasks involved.

Children will take turns, make decisions and develop their table manners. There will be a huge range of language experiences - speaking, listening to and writing orders, reading posters and menus. Numeracy will be developed through counting and price activities with money. Cutlery and chairs will need to be matched to the number of customers, meals will be sequenced and appropriate crockery

provided for each course. Science and technological development can be explored through cooking and the senses. Processes such as washing up and hygiene could be explored. Children could design and make the whole area or props to play with in it - saltdough food, plates, napkins - and can practise basic skills such as cutting and folding. They will practise balancing and carrying and will be motivated by an approach to learning in a context to which they can relate. The basic idea can be adapted to different food establishments - McDonalds, motorway services, pub restaurant. The setting could also be transferred to a different location or place in the world - such as a Parisian café!

Starting with a character

Postman Pat helps everyone in Greendale. Children can watch the videos, listen to the stories and act out Pat's role, his round and the people he meets. His van can be made out of cardboard boxes and you can re-enact some of his adventures. Pat meets others who also help people - handymen, farmers, teachers and the postmistress. Explore the journeys of parcels and letters, reinforce work on addresses and provide simple sorting and naming activities. Children will find out more about rural communities and the way people and places interconnect in a similar or different environment to their own.

Assessment

❏ Can children identify how people help us through discussion? (identification)

❏ Can they recognise pictures or photographs of people in role? (analysis)

❏ What do they use and why? Note down the vocabulary children use. (evaluation)

Planning for Knowledge and Understanding of the World: Geography and History

These pages explain how the 15 activities in this book cover all the Early Learning Goals for Knowledge and Understanding of the World. The emphasis is on activities which promote an early understanding of geography and history.

Early Learning Goals for Knowledge and Understanding of the World

KU1 Investigate objects and materials by using all of their senses as appropriate.

KU2 Find out about, and identify some features of, living things, objects and events they observe.

KU3 Look closely at similarities, differences, patterns and change.

KU4 Ask questions about why things happen and how things work.

KU5 Build and construct with a wide range of objects, selecting appropriate resources and adapting their work where necessary.

KU6 Select the tools and techniques they need to shape, assemble and join the materials they are using.

KU7 Find out about and identify the uses of everyday technology and use information and communication technology and programmable toys to support their learning.

KU8 Find out about past and present events in their own lives and in those of their families and other people they know.

KU9 Observe, find out about and identify features in the place they live and the natural world.

KU10 Begin to know about their cultures and beliefs and those of other people.

KU11 Find out about their environment and talk about those features they like and dislike.

Where we live (pages 26-27)

To be active explorers children need to develop a wide range of investigational skills linked to their senses. These activities focus on first-hand experience within the immediate locality of your setting. They will help children to establish a sense of place; identify similarities and differences between different features and places in the locality; and begin to understand that people relate to the environment in different ways.

Related goals: KU 1, 2, 3, 4, 8, 9, 11

Weather forecasting (pages 28-29)

Evidence of how the weather affects us can be explored through observation, simple recording activities and the development of a weather vocabulary. Suggestions include making a weatherboard or chart, learning popular rhymes and common sayings and recreating your own television weather studio for role play.

Related goals: KU 1, 2, 3, 4, 5, 9, 10

Going shopping (pages 30-31)

The routines associated with visits to shops and recreations of this in children's small world or role play are a natural part of many early years settings. They can also be used to extend children's knowledge and understanding of the world. The ideas here are designed to help children to develop their knowledge and understanding of different types of shops; to identify similarities and differences, patterns and reasons for different types of shops; and to understand that people relate to shops and shopping in different ways, for example as employees and customers.

Related goals: KU 1, 2, 3, 4, 8, 9, 10, 11

Outdoor play (pages 32-33)

The outdoor play area provides an opportunity to transport children to different environments, near and far. These activities encourage children to use language to describe position and location as well as the main features of the outdoor play environment; and explore the real world through outdoor recreations of places and jobs.

Related goals: KU 1, 2, 3, 4, 9, 11

What a picture! (pages 34-35)

Picture reading is an important skill. Ideas are given on how you can help children to develop this skill and the necessary vocabulary to ensure that they can explain their geographical or historical thinking.

Related goals: KU 3, 4, 7, 8, 9, 10

• • • • • • • • • • • • • • • • •

Let's have a celebration
(pages 36-37)

Birthdays, festivals, anniversaries and centenaries all provide motivating and memorable learning experiences for children and exciting ways into studying life in the past.

Related goals: KU 2, 3, 8, 10

• • • • • • • • • • • • • • • • •

Time travellers (pages 38-39)

Museums offer a wealth of opportunities, experiences and resources. Advice is given on what to do before, during and after a visit in order to get the most out of your trip.

Related goals: KU 1, 2, 3, 4, 7, 8, 10

• • • • • • • • • • • • • • • • •

The way we were (pages 40-41)

How 'The way we were' can be used as a topic to look at how children have changed and explore their developing sense of time. You can then extend this further back in time to enable children to find out more about the past and present events not only in their own lives but in those of their families and other people they know.

Related goals: KU 1, 2, 3, 4, 7, 8, 10

• • • • • • • • • • • • • • • • •

Any old iron (pages 42-43)

Starting with a collection will enable children to act as geographical, environmental or historical detectives. They will be able to discover more

about the world and how the artefacts they explore fit into it. These activities show how you help children to discover more about the world from artefacts and objects; and learn how to handle objects with care.

Related goals: KU 1, 2, 3, 4, 5, 6, 7, 8, 9, 10, 11

• • • • • • • • • • • • • • • • •

Character geography (pages 44-47)

From Postman Pat to Jack and Jill, characters appear in every rhyme, story and song that children learn. These ideas explain how they often provide a useful way into developing children's geographical understanding, helping them to find out more about the world, within and beyond their experience.

Related goals: KU 1, 2, 3, 4, 5, 6, 7, 8, 9, 10, 11

• • • • • • • • • • • • • • • • •

Jungle role play (pages 48-49)

Role-play areas can be transformed into environments associated with past times or distant places which allow your children to develop their knowledge and understanding of the world by travelling in time or to distant locations in a way that is practical and meaningful for them. So why not turn a corner of the room into a rainforest?

Related goals: KU 1, 2, 3, 4, 5, 6, 7, 8, 9, 10, 11

• • • • • • • • • • • • • • • • •

Seasons: The circle of life
(pages 50-53)

Use stories, songs, rhymes, pictures and popular television programmes to introduce and reinforce the features associated with particular seasons of

the year; and raise awareness of seasonal change in the environment.

Related goals: KU 1, 2, 3, 8, 9, 10, 11

• • • • • • • • • • • • • • • • •

Water: From raindrops to rivers (pages 54-55)

How does water move? Where does it come from? What do we use water for? What do other people use it for? Activities to help children explore an essential part of our lives.

Related goals: KU 1, 2, 3, 4, 7, 8, 9, 10, 11

• • • • • • • • • • • • • • • • •

All about me: Memories are made of this (pages 56-59)

These activities have been planned to help children to identify similarities and differences between themselves and other people; understand that people grow and change; be aware of how they relate to other people and places; and develop an 'all about me' word vocabulary.

Related goals: KU 1, 2, 3, 4, 8, 10

• • • • • • • • • • • • • • • • •

People who help us: The world of work (pages 60-61)

Young children enjoy dressing up and pretending to be other people. They encounter many different adults in their daily lives. Characters they see on children's television have specific roles. Make the most of all these opportunities to develop children's knowledge and understanding of the world of work. All the suggestions focus on learning through play.

Related goals: KU 8

Planning chart

Knowledge and Understanding of the World: Geography and History	KU 1	KU 2	KU 3	KU 4	KU 5	KU 6	KU 7	KU 8	KU 9	KU 10	KU 11
Where we live	✓	✓	✓	✓				✓	✓		✓
Weather forecasting	✓	✓	✓	✓	✓				✓	✓	
Going shopping	✓	✓	✓	✓				✓	✓	✓	✓
Outdoor play	✓	✓	✓	✓					✓		✓
What a picture!			✓	✓			✓	✓	✓	✓	
Let's have a celebration	✓	✓	✓					✓		✓	
Time travellers	✓	✓	✓	✓			✓	✓		✓	
The way we were	✓	✓	✓	✓			✓	✓			
Any old iron	✓	✓	✓	✓	✓	✓	✓	✓	✓	✓	✓
Character geography	✓	✓	✓	✓	✓	✓	✓	✓	✓	✓	✓
Jungle role play	✓	✓	✓	✓	✓	✓	✓		✓	✓	
Seasons: The circle of life	✓	✓	✓					✓	✓	✓	✓
Water: From raindrops to rivers	✓	✓	✓	✓			✓	✓	✓	✓	
All about me: Memories are made of this	✓	✓	✓	✓				✓		✓	
People who help us: The world of work								✓			